Millimeter-Wave Radar Clutter

For a complete listing of the *Artech House Radar Library*,
turn to the back of this book

Millimeter-Wave Radar Clutter

Nicholas C. Currie
Robert D. Hayes
Robert N. Trebits

Artech House
Boston • London

Library of Congress Cataloging-in-Publication Data

Currie, Nicholas C.
    Millimeter-wave radar clutter / Nicholas C. Currie, Robert D.
Hayes, Robert N. Trebits.
      p.   cm.
    Includes bibliographical references and index.
    ISBN 0-89006-345-1
    1. Radar—Interference. 2. Millimeter waves. I. Hayes R. D.
(Robert Deming), 1925- . II. Trebits, R. N. (Robert Neil),
1944- . III. Title.
TK6580.C87 1992   92-12921
621.3848—dc20   CIP

© 1992 ARTECH HOUSE, INC.
685 Canton Street
Norwood, MA 02062

International Standard Book Number: 0-89006-345-1
Library of Congress Catalog Card Number: 92-12921

10 9 8 7 6 5 4 3 2 1

Contents

Preface

Millimeter wave radars and radiometers represent the next generation of military smart sensors for detection, tracking, and surveillance. These sensors have been in development for over 40 years. The delivery of the first operational United States weapon that uses millimeter waves now appears to be at hand.

In the early 1980s, many radar engineers identified the need for reference texts emphasizing the millimeter wave radar and providing information about designing and analyzing the radars operating in this higher frequency region. In response to this need, we, along with other researchers primarily at Georgia Tech, developed the text *Principles and Applications of Millimeter Wave Radar,* published by Artech House in 1987. This handbook provides information about many aspects of millimeter wave radar systems including theory and phenomenology, components and subsystems, and applications. Each topic, however, was restricted to limited treatment in this reference handbook.

With *Millimeter-Wave Radar Clutter,* we seek to provide the radar engineer with in-depth information about both propagation and clutter backscatter effects on detection in the millimeter wave region. As a result of our significant experience in this field (including our participation in propagation/clutter backscatter measurement programs), we have access to a wide variety of millimeter wave backscatter and attenuation measurement results. We summarize this information for the reader, as well as provide additional references and reading lists.

This book covers definitions, phenomenology, theory, measured data, and detection discussion and examples. Both propagation and backscatter effects are covered, and copious data measurements are presented to support theoretical predictions. This underlying intent of this book is to provide pertinent information to assist the engineer in understanding the clutter-limited performance of a millimeter wave radar system.

The authors wish to acknowledge the understanding and patience of our families during the development and publication of this text. Special thanks are due to Ms. Phyllis Hinton for her tireless efforts in providing the superb graphics used in this text, and to Ms. Judy Truett for her hard work in integrating our separate text inputs. Finally, we would like to acknowledge Mr. Evan Chastain, Director of the Radar and Instrumentation Development Laboratory of the Georgia Tech Research Institute, for his patience and understanding, particularly during the final stages of manuscript preparation when work priorities and manuscript priorities sometimes became confused.

Nicholas C. Currie
Robert D. Hayes
Robert N. Trebits
Marietta, Georgia
March 1992

Chapter 1
Introduction

1.1 OVERVIEW

The emphasis of this book is on radar reflectivity characteristics at millimeter wavelengths. Because of Mie scattering and other scattering effects unique to the millimeter-wave region, it is not possible to simply scale X-band radar reflectivity data to 35 or 95 GHz and expect to achieve a realistic approximation of the actual situation. Thus, the material covered by the authors addresses those particular characteristics of clutter returns in the millimeter-wave region that are important to the radar systems engineer in the design and understanding of millimeter-wave sensors, particularly radars.

This chapter addresses the fundamental physical and mathematical definitions of radar cross section (RCS) and radar reflectivity. The concept of RCS is extended to polarization and frequency domains and to spatial effects, in addition to the usual time domain—amplitude characterization. Volume and surface scattering concepts are discussed, with respect to normalized radar reflectivity. Attenuation concepts from the radar systems engineer's perspective are also covered. The final section discusses how to effectively use the information summarized in this book.

1.2 WHAT IS RADAR CLUTTER?

Before starting out a discussion of radar clutter characteristics, it is reasonable to ask the question "what is clutter?" The answer to such a question can be either very simple or very complex. H.A. Corriher, Jr., has answered this question by defining radar clutter at its most basic level [1]:

> *Clutter is a return or a group of returns that is unwanted in the radar situation being considered.*

This statement leads to the following consequence:

One radar's clutter is another radar's target!

If this definition is accepted, then clutter can, for certain situations, be almost any collection of objects that reflect energy back to the radar. However, the majority of the time the target to be detected is manmade, and the interfering returns to the radar are from naturally occurring objects. Thus, the focus of this book will be on the millimeter-wave radar reflectivity of naturally occurring objects. In addition, information will be included for manmade smoke and obscurant material intended to defeat millimeter-wave sensors.

1.3 RADAR CROSS SECTION DEFINITIONS

The term radar cross section will be defined in terms of (1) a transfer function parameter expressing the relationship between the magnitudes of the transmitted signal and the received radar signal from the target and (2) the "far field" limit of the ratio of signal power at the radar receiver to that incident at a target. Polarization-dependent parameters and their relationship to RCS will be defined to completely describe all of the reflective properties of an object. Both surface and volume definitions for the radar resolution cell are included to define the RCS for each particular geometry situation.

1.3.1 The RCS Concept

The concept of the RCS σ of an object is the apparent "size" of the object as observed by radar. For the monostatic case, where the transmitter and receiver antenna are one and the same, the RCS is a measure of backscattered power normalized with respect to the power density of the radar signal incident on the object. Furthermore, the RCS of an object is defined to be a characteristic of the object alone and is not dependent on the range between the object and the radar.

The power density at the object location (target) at a range R from a directive antenna of gain G_t is

$$P_{tar} = \frac{P_t G_t}{4\pi R^2} \qquad (1.1)$$

where P_t is the transmitted power. The power density P_a at the receiving antenna of the electromagnetic signal re-radiated by the object will then be

$$P_a = \left(\frac{P_t G_t}{4\pi R^2}\right)\left(\frac{\sigma}{4\pi R^2}\right) \qquad (1.2)$$

where σ represents the RCS of the object.

Note that the RCS defined above is a function of the radar signal wavelength and polarization, the physical composition and shape of the object, and the orientation of the object with respect to the radar line of sight. It is also implicitly assumed that the radar resolution cell is large; that is, the azimuthal extent of the illuminated spot $(R\Theta_{az})$ is larger than the width of the target, and the pulse length τ is long (i.e., $\tau \gg 2L/c$, where L is the range extent of the target and c is the speed of light). Otherwise, only part of the object will be instantaneously illuminated, and the backscattered signal will be the vector sum of reflections from parts of the object rather than the entire object. A fully illuminated target is known as a *point* target.

The received power P_r captured by the receiver antenna having an effective aperture A_e is

$$P_r = \left(\frac{P_t G_t}{4\pi R^2}\right)\left(\frac{\sigma}{4\pi R^2}\right) A_e \tag{1.3}$$

The receiver antenna gain G_r can be expressed in terms of the antenna's effective aperture A_e by means of the relationship

$$G_r = \left(\frac{4\pi A_e}{\lambda^2}\right) \tag{1.4}$$

where λ is the radar wavelength. Finally, the received power P_r can be expressed as

$$P_r = \left(\frac{P_t G_t}{4\pi R^2}\right)\left(\frac{\sigma}{4\pi R^2}\right)\left(\frac{G_r \lambda^2}{4\pi}\right) \tag{1.5}$$

or

$$P_r = \left(\frac{P_t G^2 \sigma \lambda^2}{(4\pi)^3 R^4}\right) \tag{1.6}$$

where the radar transmitter and the receiver antennas are the same; that is, $G_t = G_r = G$.

It is seen from Equation (1.6) that the RCS parameter σ quantitatively relates the transmitter power to the power backscattered by the object toward the radar, and to the power captured by the receiver antenna. Equation (1.6) represents the simplest form of the *radar range equation.*

The fact that an object's RCS is a characteristic of an object and not its environment implies that the object is in the *far field* of the radar antenna, and is not dependent on range. This requirement is equivalent to the object being physically far removed from the radar antenna such that the electromagnetic wave fronts incident on the object are, for all practical purposes, planar. This fact is evident in an equivalent theoretical definition of RCS:

$$\sigma = \lim_{R\to\infty} 4\pi R^2 \left|\frac{E_r^2}{E_i^2}\right| \tag{1.7}$$

where

E_r = electric field magnitude at the radar receiver
E_i = electric field magnitude incident at the object

Alternatively, the RCS can be defined in terms of the analogous magnetic field parameters:

$$\sigma = \lim_{R\to\infty} 4\pi R^2 \left|\frac{H_r^2}{H_i^2}\right| \tag{1.8}$$

where

H_r = magnetic field magnitude at the radar receiver
H_i = magnetic field magnitude incident at the object

The radar will in fact be located at a finite distance from an object. How then does one define the far field distance between the radar and the object whose cross section is in consideration? From a heuristic point of view, the far field distance from the radar antenna can be defined in terms of the deviation of the phase front from perfect uniformity over an aperture equal to the appropriate cross-range dimension of the antenna (Knott [2]). It is straightforward to show that the one-way phase deviation ϕ, in radians, is given by the expression

$$\phi \cong \frac{\pi d^2}{4R\lambda} \tag{1.9}$$

where d is the largest cross-range dimension of the antenna. Equation (1.9) represents the phase difference of signals emanating from the center of the antenna and its extremity at a specific target location. If the maximum allowable one-way phase deviation is arbitrarily defined to be $\pi/8$ rad (22.5°), then the far field condition is achieved for ranges R given by

$$R \geq \frac{2d^2}{\lambda} \tag{1.10}$$

The phase deviation above corresponds to a 1 dB amplitude taper across the target. Note that it is also possible to define the far field from the target in an analogous manner, considering the maximum "width" of the object as its aperture. As a result, many large targets may have a "far field" distance measured in kilometers.

1.3.2 RCS Polarization Dependence

The polarization dependency of an object's RCS may be characterized using a mathematical description based on the relationship between the transmitted and received electric fields in terms of an orthogonal polarization basis set. For example, one such basis set consists of the linear polarizations represented by E_h and E_v, where h denotes horizontal polarization, and v denotes vertical polarization.

The received electric field E_r has two orthogonal polarization components defined by the vector basis set $\hat{x}$ and $\hat{y}$, and can be mathematically described as

$$E_r = E_x \cos \Phi_x \, \hat{x} + E_y \cos \Phi_y \, \hat{y} \qquad (1.11)$$

where subscripts x and y respectively refer to amplitudes E and phases Φ, in the x and y directions, respectively, where

$$\Phi_x = \omega t - kR + \phi_x$$
$$\Phi_y = \omega t - kR + \phi_y$$

and

ω = radian frequency
t = time
k = wave number (= $2\pi/\lambda$)
λ = wavelength
R = range
ϕ_x = x-component of initial signal phase
ϕ_y = y-component of initial signal phase

With no loss in generality, Equation (1.11) can be written as

$$E_r = E_x \, \hat{x} + E_y \cos \Phi \, \hat{y} \qquad (1.12)$$

where the x-component phase has been arbitrarily set to zero, and the frequency and range components of phase have been suppressed. Equation (1.12) can be expressed in terms of orthogonal parameters x and y, yielding the expression

$$\left[\frac{E_y^2}{E_x^2}\right] x^2 - \left[\frac{2E_y \cos \phi}{E_x}\right] xy + y^2 - E_y^2 \sin^2 \phi = 0 \qquad (1.13)$$

Equation (1.13) represents, in this case, an ellipse whose semimajor axis has been rotated through an angle Θ given by the equation

$$\cot 2\Theta = \frac{[(E_y/E_x)^2 - 1]}{2(E_y/E_x) \cos \phi} \qquad (1.14)$$

Equation (1.13) defines the *polarization ellipse,* shown in Figure 1.1, which can be graphically described by its orientation angle Θ (with respect to horizontal and vertical axes) and its ellipticity angle τ (or ellipticity ratio r or eccentricity e).

The ellipticity angle τ is defined by the equation

$$\tau = \tan^{-1}\left[\frac{E_y}{E_x}\right], \quad |\tau| \leq 45° \qquad (1.15)$$

Alternatively, the ellipticity ratio r is defined by the equation

$$r = \tan \tau = \frac{E_y}{E_x}, \quad |r| \leq 1 \qquad (1.16)$$

and the eccentricity e is defined by the equation

$$e = \sqrt{1 - E_y^2/E_x^2} = \sqrt{1 - r^2}, \quad e \leq 1 \qquad (1.17)$$

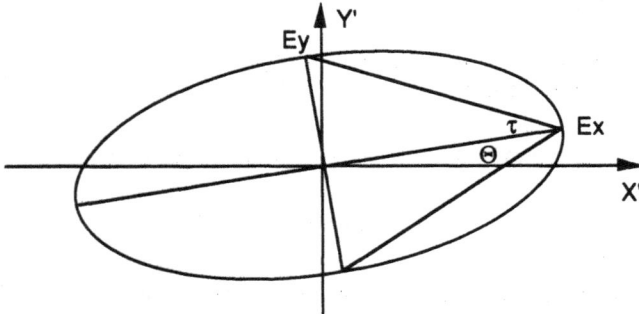

Figure 1.1 Polarization ellipse.

The received electric field vectors E_r may then be related to the transmitted electric field vectors E_t by the pair of equations

$$E_h^r = a_{hh} E_h^t + a_{vh} E_v^t \tag{1.18}$$

$$E_v^r = a_{hv} E_h^t + a_{vv} E_v^t \tag{1.19}$$

or in matrix notation,

$$\begin{bmatrix} E_h^r \\ E_v^r \end{bmatrix} = \begin{bmatrix} a_{hh} & a_{vh} \\ a_{hv} & a_{vv} \end{bmatrix} \begin{bmatrix} E_h^t \\ E_v^t \end{bmatrix} \tag{1.20}$$

where h denotes one polarization (say horizontal) and v denotes the orthogonal polarization (say vertical). A similar matrix expression can be expressed for a circular polarization basis set.

The matrix $\begin{bmatrix} a_{hh} & a_{vh} \\ a_{hv} & a_{vv} \end{bmatrix}$ is the full polarization target matrix and contains all the information backscattered from a target for a particular target orientation with respect to the radar line of sight, a specific radar frequency, and a specific polarization basis set. The parameters a_{xx} are complex valued; the first subscript denotes the transmitted polarization, while the second polarization denotes the received polarization. Thus, a_{vh} is a complex-valued coupling parameter between the vertically polarized transmission signal and the horizontally polarized received signal. The latter equation is called the *full polarization scattering matrix* for the target and contains all the information that exists about the scattering properties of that target at that particular aspect with respect to the radar line of sight.

The parameters a_{xy} can be related to the appropriate RCS parameters via the expression

$$a_{xy} = \sqrt{\sigma_{xy}}\, e^{j\varphi_{xy}} \tag{1.21}$$

so that the full polarization target matrix becomes

$$\begin{bmatrix} a_{hh} & a_{vh} \\ a_{hv} & a_{vv} \end{bmatrix} = \begin{bmatrix} \sqrt{\sigma_{hh}}\, e^{j\varphi_{hh}} & \sqrt{\sigma_{vh}}\, e^{j\varphi_{vh}} \\ \sqrt{\sigma_{hv}}\, e^{j\varphi_{hv}} & \sqrt{\sigma_{vv}}\, e^{j\varphi_{vv}} \end{bmatrix} \tag{1.22}$$

For a monostatic radar configuration, where the transmit and receive antennas are the same, a_{hv} will equal a_{vh}. Furthermore, one of the four phase parameters φ is arbitrary; by convention, φ_{hh} is set to zero. Thus, there are only five independent parameters in the full polarization target matrix, reducing the matrix to

$$\begin{bmatrix} a_{hh} \ a_{vh} \\ a_{hv} \ a_{vv} \end{bmatrix} = \begin{bmatrix} \sqrt{\sigma_{hh}} & \sqrt{\sigma_{vh}} \ e^{j\varphi_{vh}} \\ \sqrt{\sigma_{vh}} \ e^{j\varphi_{vh}} & \sqrt{\sigma_{vv}} \ e^{j\varphi_{vv}} \end{bmatrix} \qquad (1.23)$$

The expression for circular polarizations equivalent to Equation (1.20) is

$$\begin{bmatrix} E^r_1 \\ E^r_2 \end{bmatrix} = \begin{bmatrix} c_{11} & c_{21} \\ c_{12} & c_{22} \end{bmatrix} \begin{bmatrix} E^i_1 \\ E^i_2 \end{bmatrix} \qquad (1.24)$$

where the matrix elements c_{ij} are the circular polarization analogs to the previous elements a_{ij}. For backscatter geometries, it is also true, by reciprocity, that $c_{12} = c_{21}$. It is straightforward to derive the following relationships between the a_{ij} and c_{ij} elements (Long, [3]):

$$|c_{11}| = \left| \frac{a_{hh} - a_{vv}}{2} + ja_{hv} \right| \qquad (1.25)$$

$$|c_{12}| = |c_{21}| = \left| \frac{a_{hh} + a_{vv}}{2} \right| \qquad (1.26)$$

$$|c_{22}| = \left| \frac{a_{hh} - a_{vv}}{2} - ja_{hv} \right| \qquad (1.27)$$

1.3.3 Volume RCS Definitions

Millimeter-wave signal backscatter from atmospheric constituents is conventionally quantified by the parameter η, which represents the RCS per unit volume of the scattering material. The volume of the resolution cell is defined by antenna patterns in azimuth and in elevation, and by the range resolution $c\tau/2$, where τ is the pulse length in seconds, τ is the radar pulse length, and c is the velocity of light (3×10^8 m/s). For radars that employ pulse compression techniques, τ is the compressed pulse length. For radars that employ frequency modulated constant wave (FMCW), τ is approximately the reciprocal of the total effective bandwidth. As shown in Figure 1.2, if the radar antenna has a pencil beam pattern, the azimuth and elevation beamwidths, defined by the 3-dB points of the beam pattern, will determine the extent of the resolution cell in the vertical and horizontal dimensions. If the azimuth and elevation beamwidths are given by θ_{az} and θ_{el}, then the resolution cell cylinder will have diameters of just $R\theta_{az}$ and $R\theta_{el}$, respectively, where R is the range to the resolution cell. The range extent of the resolution cell will be simply $c\tau/2$.

Then the volume V of the cell will be the area of the cell cross section $\pi(R\theta_{az})(R\theta_{el})/4$ times the length $c\tau/2$ of the cell:

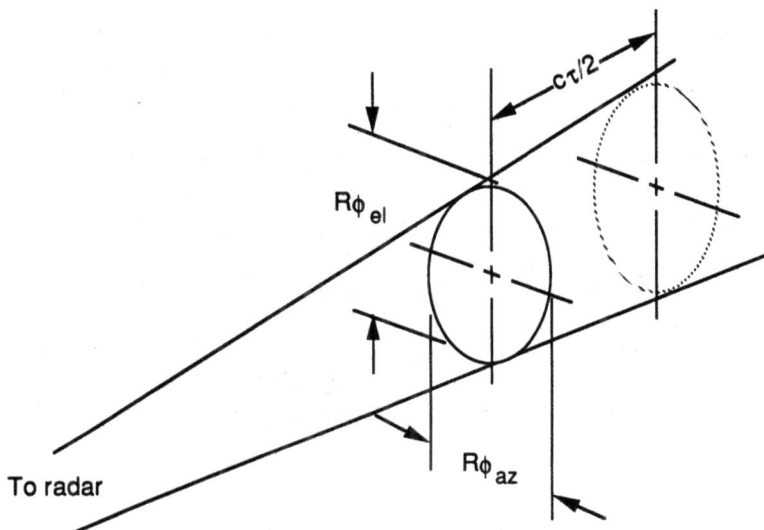

Figure 1.2 Volumetric resolution cell for a pencil beam antenna pattern. (From Trebits [4].)

$$V = \left[\frac{\pi R^2 \Theta_{el} \Theta_{az}}{4} \right] \times \left[\frac{c\tau}{2} \right] \tag{1.28}$$

A beam shape factor of $1/\alpha^2 = 1/1.77$ is usually used to reduce the calculated volume V. This factor arises from a one-way factor of $1/1.33$ due to a Gaussian beam shape rather than a rectangular beam shape. The value of the normalized backscatter parameter η is then just $(\sigma/V\alpha^2)$, where σ is the actual *measured* RCS of the clutter within the resolution volume.

1.3.4 Surface RCS Definitions

Land and sea surface backscatter is characterized by σ^0, the radar reflectivity per unit area illuminated by the radar, or

$$\sigma^0 = \sigma/A \tag{1.29}$$

where σ is the RCS of the illuminated area and A is the illuminated area of the sea surface or ground. By convention, the reflectivity σ^0 is shown in superscript form and is usually specified as *sigma zero* or *sigma naught*. However, surface reflectivity will occasionally be expressed in the literature in terms of the parameter γ, which is essentially independent of the incident angle Θ to the surface.

$$\gamma = \sigma^0/\sin \Theta \tag{1.30}$$

The parameter σ^0 is a complex function of the radar's frequency and polarization, plus the surface's roughness, tilt variation, content, and the geometry between the radar line of sight and the local surface normal. For example, ground reflectivity depends on the soil makeup, water content, roughness, tilt variation, snow cover, and foliage cover, among other factors. Sea reflectivity will depend on the average wave height, wind speed and direction, and wave direction.

The computation of the area A on the sea or ground surface, which is instantaneously illuminated by the radar, will depend on the geometry of the radar line of sight, elevation beamwidth, pulse length, and incidence angle to the local surface. Two different geometries are defined: the pulse length limited case and the beamwidth limited case. The pulse length limited case is defined where the radar's range resolution, projected onto the surface, is smaller than the range extent of the area on the surface instantaneously illuminated by the antenna beam, as shown in Figure 1.3, (Trebits [5]).

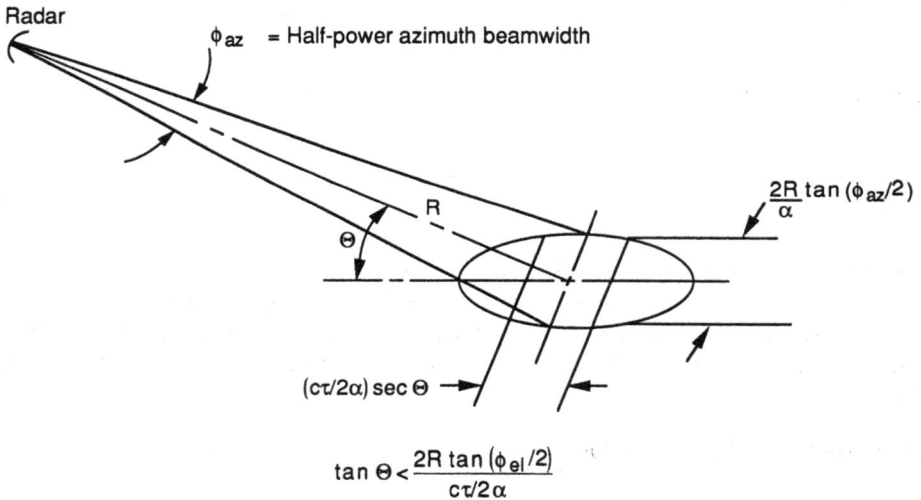

Figure 1.3 Pulse length limited case geometry.

For this geometry, the instantaneously illuminated resolution cell is approximately rectangular and can be calculated by multiplying the projected range resolution $(c\tau/2)\sec \Theta$ by the width of the of the illuminated ellipse $2R \tan(\phi_{az}/2)$. The one-way azimuth beamwidth is ϕ_{az}, and Θ is the incidence angle to the local surface tangent. A beam shape factor $1/\alpha$ of $1/1.33$ for a Gaussian beam shape may be included to account for nonuniform beamwidth illumination. The area A for the pulse length limited case is then given by

$$A = 2R(c\tau/2\alpha) \tan(\phi_{az}/2) \sec \Theta \qquad (1.31)$$

where

$$\tan \Theta < \frac{2R \tan(\phi_{el}/2)}{c\tau/2\alpha} \qquad (1.32)$$

The small angle ($\Theta < 10°$) approximations for Equations (1.31) and (1.32) are

$$A = R(c\tau/2\alpha)\phi_{az} \sec \Theta \qquad (1.33)$$

and

$$\tan \Theta < \frac{2R \tan(\phi_{el}/2)}{(c\tau/2\alpha)} \qquad (1.34)$$

The beam limited case is defined for a geometry where the radar range resolution projected onto the ground or sea surface is larger than the range extent of the area instantaneously illuminated by the radar beam, as demonstrated in Figure 1.4 (Trebits [6]).

The resolution cell in this case is elliptical, where the range axis diameter is given by

$$\frac{2R}{\alpha} \tan(\phi_{el}/2) \csc \Theta \qquad (1.35)$$

where ϕ_{el} is the one-way antenna elevation beamwidth, and the azimuth axis diameter is given by

$$\frac{2R}{\alpha} \tan(\phi_{az}/2) \qquad (1.36)$$

where ϕ_{az} is the one-way antenna azimuth beamwidth. The $1/\alpha$ beam shape factor has been applied to both axes.

The area A of the resolution cell for the beamwidth limited case is thus the area of an ellipse with the diameters given by Equations (1.35) and (1.36):

$$A = \frac{\pi R^2}{\alpha^2} \tan(\phi_{az}/2) \tan(\phi_{el}/2) \csc \Theta \qquad (1.37)$$

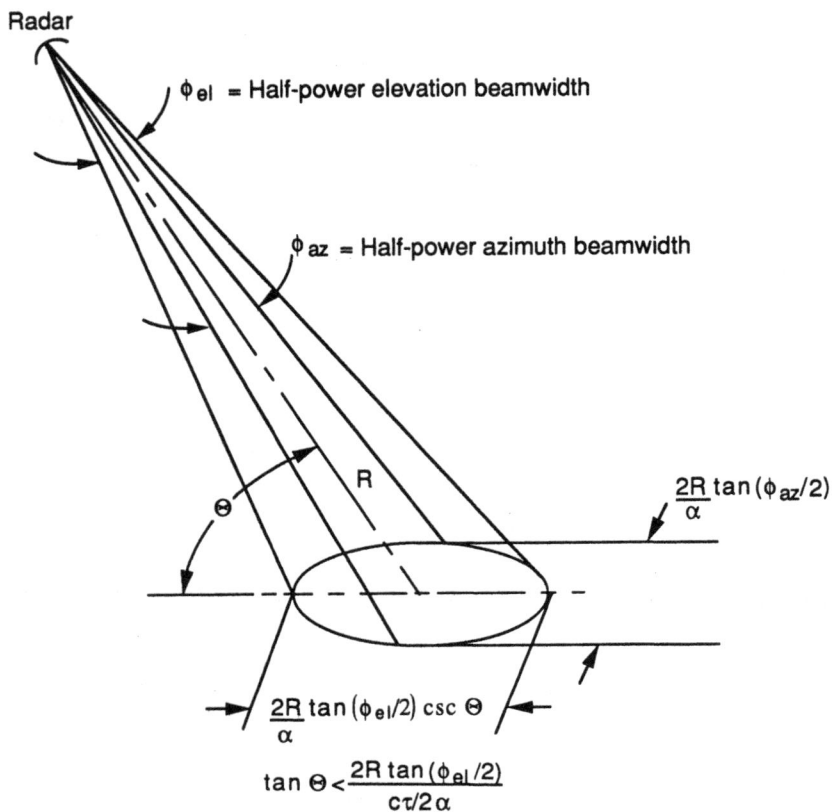

Figure 1.4 Beam limited case geometry. (After F.E. Nathanson, *Radar Design Principles*, McGraw Hill Book Company, New York, 1969, p. 65.)

where

$$\tan \Theta > \frac{2R \tan(\phi_{el}/2)}{(c\tau/2\alpha)} \tag{1.38}$$

The small angle approximations (Θ_{az}, $\Theta_{el} < 10°$) for Equations (1.37) and (1.38) are

$$A = \frac{\pi R^2}{4\alpha^2} \phi_{az}\phi_{el} \csc \Theta \tag{1.39}$$

and

$$\tan \Theta > \frac{R\phi_{el}}{c\tau/2\alpha} \tag{1.40}$$

1.4 ATTENUATION DEFINITIONS

The radar signal will suffer losses as it propagates through the air between the radar and the target/clutter cells and back again. These losses will be caused by interactions with the air molecules themselves and by any particles in the air along the radar line of sight, such as rain, snow, dust, or explosive ejecta. The loss in signal strength due to the particles can be attributed to two different physical processes: the absorption of signal energy (transformation to heat) and the reradiation of signal energy over the entire 4π steradians (forward scattering and backscattering). The sum of the two effects represents the total loss of signal energy and is termed *signal extinction*. This terminology is taken from the optical and infrared research community and differs from the radar engineering term *signal attenuation*, which does not differentiate by the signal loss process. These different effects are characterized by appropriate cross sections:

$$Q_s = \frac{\text{Total power scattered}}{\text{Incident power density}}$$

$$Q_a = \frac{\text{Total power absorbed}}{\text{Incident power density}}$$ (1.41)

$$Q_e = \frac{\text{Total power lost}}{\text{Incident power density}}$$

where Q_s, Q_a, and Q_e are the scattering, absorption, and extinction cross sections, respectively.

For consistency, the nomenclature of the radar engineering community will be used, where the appropriate terminology is signal attenuation coefficient, and the conventional symbol is α (in decibels per meter). When signal attenuation due to both absorption and to scattering processes is added to Equation 1.6, the expression for received power becomes

$$P_r = \left[\frac{P_i G^2 \sigma \lambda^2 e^{-2R\alpha'}}{(4\pi)^3 R^4} \right]$$ (1.42)

where the one-way signal attenuation coefficient α', expressed in nepers per meter (Np/m), is assumed to be constant over the entire range extent. When the attenuation is not uniform over the entire range extent, the total two-way attenuation L is found by integrating over the range extent of the lossy medium:

$$L = \exp\left(-4 \int_{R_{\text{near}}}^{R_{\text{far}}} \alpha'(R)dR \right)$$ (1.43)

Conventionally, Equation (1.42) is used by the radar systems engineer in logarithmic form, and the medium loss $10 \log_{10} L$, in decibel (dB) units, is expressed as

$$10 \log_{10} L = 10 \left| \log_{10}[e^{-2R\alpha'}] \right| \tag{1.44}$$

$$10 \log_{10} L = 10(-2R\alpha') \log_{10}(e) \tag{1.45}$$

$$10 \log_{10} L = 10(-2R\alpha')(0.4343) = -8.686R\alpha' = -2R\alpha \tag{1.46}$$

where the attenuation coefficient α is in dB/unit distance.

1.5 HOW DO WE DESCRIBE CLUTTER?

We describe clutter in terms of measurable quantities that have an impact on the detectability of a target in clutter. The quantities include amplitude properties, spectral/temporal varying characteristics, and spatial varying properties. Amplitude properties include describing the values of amplitudes for all the scatterers within the clutter resolution cell and how they add vectorially to form a composite signal. Temporal and spectral properties describe how the composite signal from clutter changes. Spatial properties describe how the clutter varies in space as a radar beam moves across the clutter environment.

The radar signal amplitude, polarization, and phase scattered from a resolution cell containing a spatially distributed target will fluctuate because of relative motion among the constituent components, or change in aspect angle because of motion of the radar relative to the resolution cell. For example, wind will move leaves and grass on land or create waves and riplets on the sea surface. Since the backscattered signal entering the radar receiver antenna is the vector sum of returns from each scattering object within the resolution cell, any relative movement among those separate scatterers can drastically affect the vector sum. The *amplitude probability distribution function* of the radar signal is often used to calculate target detection rates in an environment of receiver noise and background clutter competition. The *frequency spectrum* of the radar signal is often used to determine cutoff frequencies for moving-target indicator (MTI) processors. The *autocorrelation function* of the radar signal is often used to indicate optimum integration times for target detection.

1.5.1 Signal Amplitude Properties

Clutter backscatter probability distribution functions are a more realistic way to define clutter, although they are more difficult to measure. The clutter backscatter histogram is a signal distribution that describes how often the clutter power has a certain value, as shown in Figure 1.5. If the vertical axis is normalized by dividing the

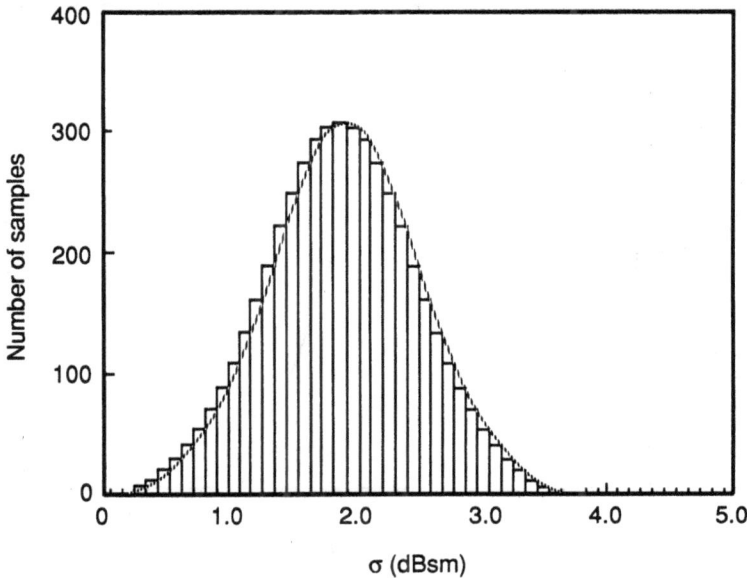

Figure 1.5 Histogram of a number of radar clutter backscatter samples versus clutter backscatter power.

number of specific values by the total number of data points, then the histogram represents a density function and describes the probability that the clutter backscatter power will have a specific value, as shown in Figure 1.6. Finally, if the area under the density curve and horizontal axis is calculated from negative infinity to a specific data value, a cumulative distribution is obtained representing the cumulative probability that the clutter power exceeds this specific value, as shown in Figure 1.7.

Rivers [7] developed a useful rule of thumb for most distribution types. The following relationship holds:

$$\bar{\sigma} = \sigma_{90\%} - 3.5 \text{ dB} \pm 0.5 \text{ dB} \tag{1.47}$$

where

$\bar{\sigma}$ = mean value of clutter

$\sigma_{90\%}$ = 90% point of the distribution

The statistical characterizations of a parameter of a radar signal are derived from the probability density function P for that function. The probability density function $W(X)$ of the "random variable" X is defined such that if X is between X and $X + dX$, then

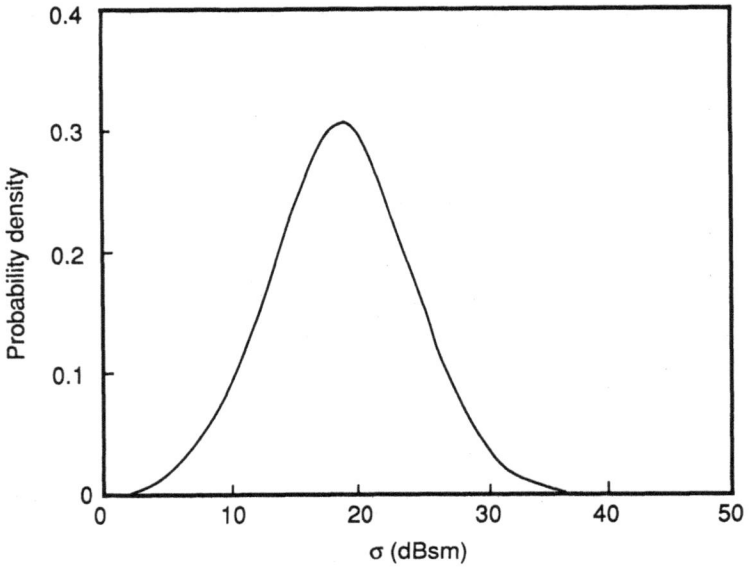

Figure 1.6 Probability density function (pdf) for the histogram of Figure 1.5.

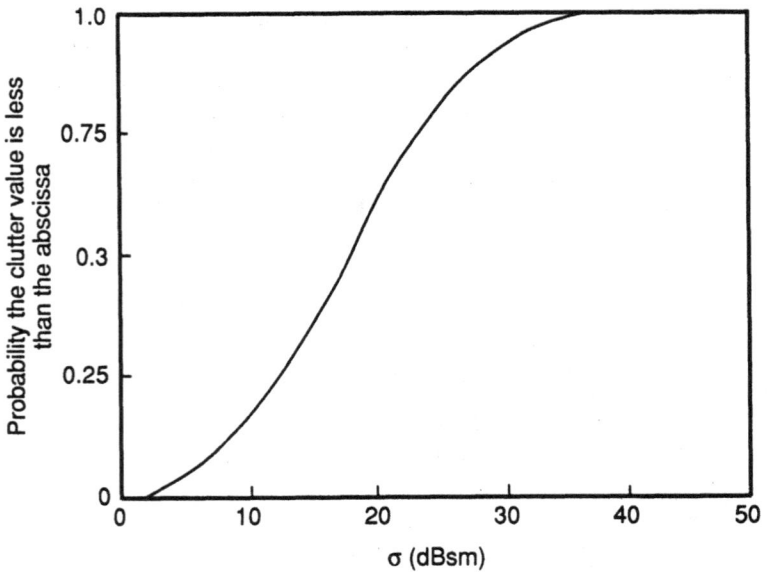

Figure 1.7 Cumulative probability density function created from the density function of Figure 1.6.

$$W(X) = P\{X \text{ between } X \text{ and } X + dX\} \tag{1.48}$$

and

$$P\{X_1 \leq X \leq X_2\} = \int_{X_1}^{X_2} W(X)dX \tag{1.49}$$

Equation (1.49), integrated over all real numbers, must yield unity, which is equivalent to the fact that the "area" under the probability distribution curve must also equal unity:

$$\int_{-\infty}^{\infty} W(X)dX = 1 \tag{1.50}$$

The average, or mean value of the variable X is denoted by the overbar symbol, as in $\bar{X}$, by μ, or as the expectation value $\langle X \rangle$:

$$\bar{X} = \langle X \rangle = \int_{-\infty}^{\infty} XW(X)dX \tag{1.51}$$

The *time average* or *temporal mean* of X over the interval T is given by the expression

$$\bar{X} = \frac{1}{T} \int_0^T X(t)dt \tag{1.52}$$

The median value and the 90% (peak) value are obtained from the cumulative distribution function. Note that if clutter data are to be used for target signal-to-clutter estimation, then the *mean* value of the clutter should be used rather than the median, yielding the mean signal-to-clutter ratio. Target detectability then becomes a matter of the target model and the clutter variance. For a nonsymmetric distribution, such as that shown in Figure 1.8, the mean value is always greater than the median value.

The variance S^2 of the variable X represents the "spread" or range of fluctuation of the data from the mean value, and is defined by the expression

$$S^2 = \langle (X - \bar{X})^2 \rangle = \int_{-\infty}^{\infty} (X - \bar{X})^2 W(X)dX \tag{1.53}$$

and can be seen to be just the average of the squared deviations from the mean of

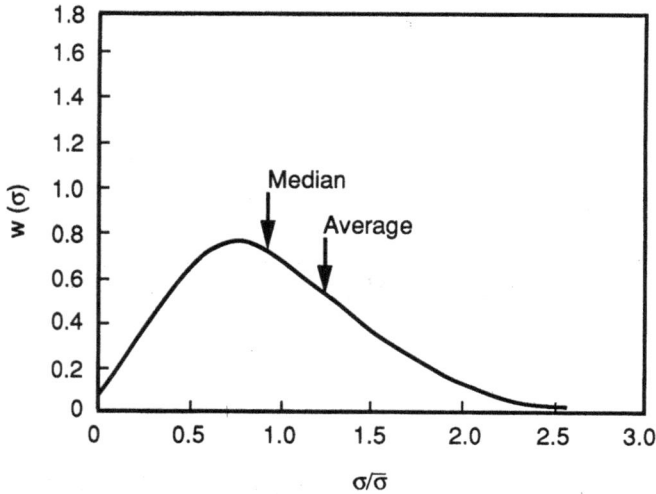

Figure 1.8 Median and mean values for an asymmetric probability density function.

X. The *standard deviation* S (also denoted by σ in the literature) is the square root of the variance and has the same units as the variable X:

$$S = \left[\int_{-\infty}^{\infty} (X - \bar{X})^2 W(X)dX \right]^{1/2} \tag{1.54}$$

The median X_m of the variable X is the "middle value" of X, that value of X for which X is less than exactly one-half of the time, or mathematically,

$$P\{X \le X_m\} = 0.5 = \int_0^{X_m} W(X)dX \tag{1.55}$$

Finally, the mode of the variable X is that value of X for which the probability amplitude distribution function is greatest. Figure 1.9 shows a representative probability amplitude distribution function, along with its mean, median, and mode.

The most commonly used amplitude probability distribution functions that characterize RCS of fluctuating and distributed targets include the *Gaussian, Rayleigh, Ricean, lognormal,* and *Weibull* functions. The Gaussian, or *normal*, amplitude probability distribution results from many scattering centers. The Gaussian probability amplitude distribution function is mathematically characterized for a given $W(X)$, by constant values for S and $\bar{X}$; $S > 0$ and $\bar{X} = X_m$. The Gaussian function

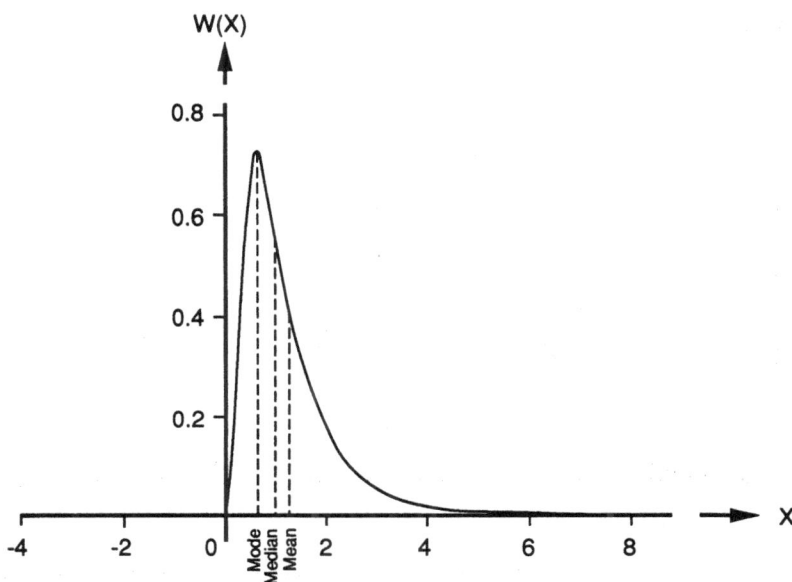

Figure 1.9 Representative amplitude probability distribution function showing mean, median, and mode. (From Aitcheson and Brown [8], © 1969 Cambridge University Press.)

also has the property that $\langle X^2 \rangle = S^2$ if $\bar{X} = 0$. Mathematically, the Gaussian amplitude probability distribution function is given by the expression

$$W(X) = \frac{1}{S\sqrt{2\pi}} \exp\left[\frac{1}{2}\left(\frac{X - \bar{X}}{S}\right)^2\right] \tag{1.56}$$

Figure 1.10 shows a Gaussian amplitude probability distribution function, where $SW(X)$ is plotted against X/S and $\bar{X} = 0$, making the curve independent of S. Sixty-eight and ninety-five percent of the data points are within one and two standard deviations of the median, respectively (Long [9]).

The Rayleigh probability distribution, plotted in Figure 1.11, is the result of the noncoherent scattering from a reasonable number of scatterers approximately equal in size, whose returns are randomly distributed in phase. The Rayleigh power probability distribution function results from detection by a square law detector and is characterized by $X \geq 0$, $\alpha > 0$, where α is a constant for a given $W(X)$. Mathematically, the Rayleigh power probability distribution function is given by the expression

$$W(X) = \frac{1}{2S^2} e^{-\left(\frac{X}{2S^2}\right)} \tag{1.57}$$

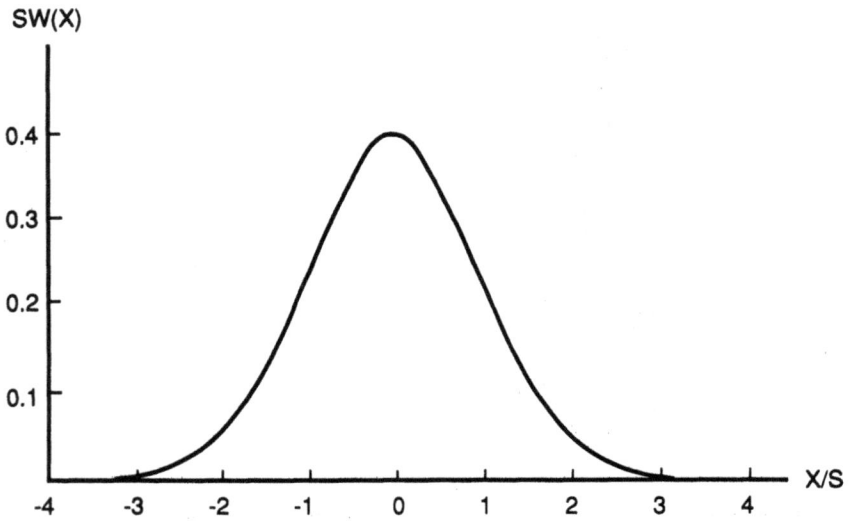

Figure 1.10 The Gaussian amplitude probability distribution function. (From Long [9], © 1984 Artech House.)

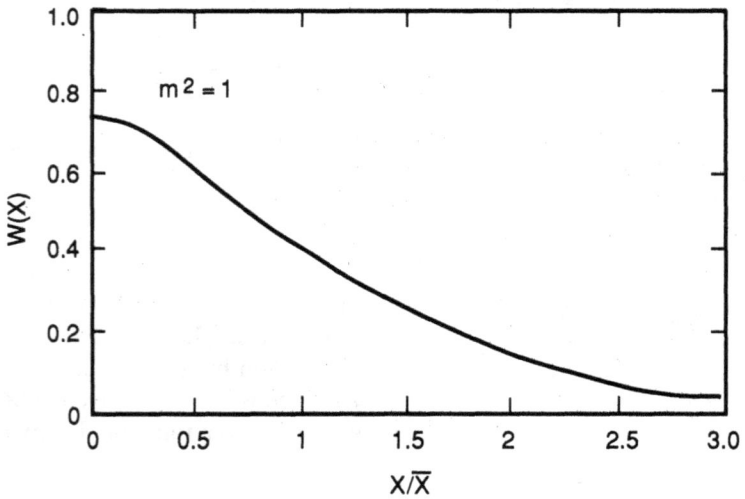

Figure 1.11 The Rayleigh power probability distribution function.

where S is the standard deviation. Figure 1.11 shows a graph of a Rayleigh power probability distribution function.

The Ricean amplitude probability distribution function is characterized by a constant-amplitude dominant scatterer and a large number of Rayleigh distributed scatterers. Mathematically, the Ricean amplitude probability distribution function is given by the expression

$$W(X) = \frac{1 + m^2}{\bar{X}} \exp\left[-m^2 - \frac{X}{\bar{X}}(1 + m^2)\right] I_0\left(2m\sqrt{1 + m^2\left(\frac{X}{\bar{X}}\right)}\right) \quad (1.58)$$

where the modified Bessel function I_0 is given by

$$I_0(Y) = 1 + \frac{Y^2}{2^2} + \frac{Y^4}{2^2 \times 4^2} + \frac{Y^6}{2^2 \times 4^2 \times 6^2} + \cdots \quad (1.59)$$

and m^2 is the ratio of the amplitude of the constant component to the average amplitude of the radar signal. For large values of m, the Ricean distribution approaches a Gaussian shape with a nonzero mean.

A representative Ricean amplitude probability distribution function is shown in Figure 1.12.

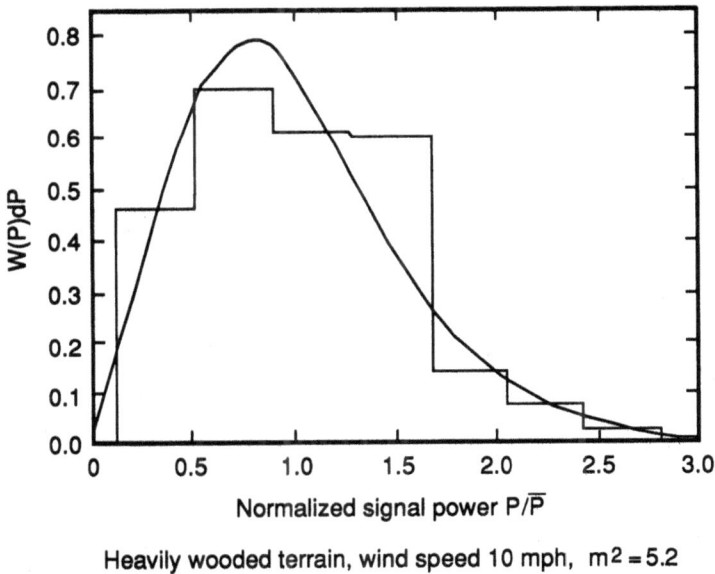

Heavily wooded terrain, wind speed 10 mph, $m^2 = 5.2$

Figure 1.12 The Ricean amplitude probability distribution function. (From Kerr [10], © 1951 McGraw Hill Book Company.)

22

The lognormal amplitude probability distribution is the result of scattering from a collection of scatterers of different sizes, including a small number of large scatterers, whose backscatter phase is randomly distributed. The lognormal amplitude probability distribution function is mathematically characterized by a variable whose logarithm values are normally distributed, as expressed in the expression

$$W(X) = \frac{1}{XS\sqrt{2\pi}} \exp\left[-\frac{(\ln X - \bar{X})^2}{2S^2}\right] \qquad (1.60)$$

where

$\bar{X} = \langle \ln X \rangle$
$S^2 = $ variance of $\ln X$

A representative lognormal amplitude probability distribution function is shown in Figure 1.13.

The Weibull amplitude probability distribution is the result of scattering from a collection of several size classes of scatterers. The Weibull amplitude probability distribution function is mathematically described by the expression

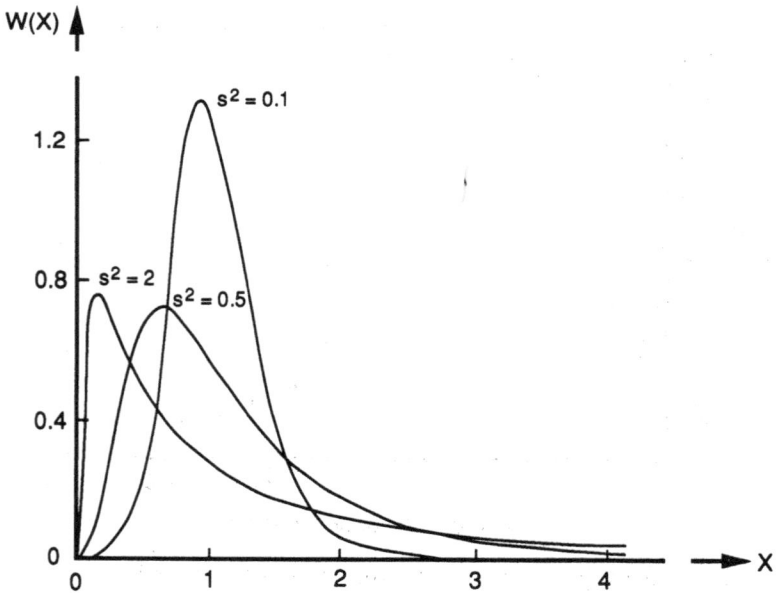

Figure 1.13 The lognormal amplitude probability distribution function. (From Aitcheson and Brown [11], © 1969 Cambridge University Press.)

$$W(X) = \frac{bX^{b-1}}{\alpha} \exp\left[\frac{X^b}{\alpha}\right] \qquad (1.61)$$

where b is a fitting parameter, $\alpha = (Xm^b/\ln(2))$.

A representative Weibull amplitude probability distribution function is shown in Figure 1.14.

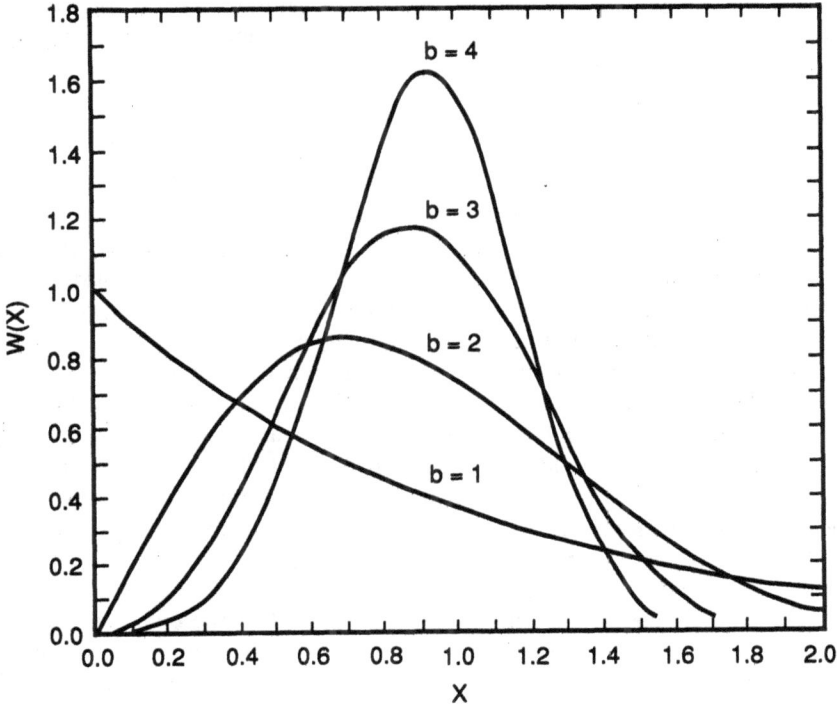

Figure 1.14 The Weibull amplitude probability distribution function.

1.5.2 Spectral and Temporal Properties

The frequency content of a radar backscatter signal contains information that can be used for (1) filter shaping in moving target indication processors or rain rejection filters, (2) estimating closing speed between the radar platform and a target, and (3) classifying or identifying targets based on spectral fine structure characteristics. The *Fourier transform* of the time domain radar backscatter signal is usually used to extract frequency domain or spectral information. If the voltage versus time function of the radar backscatter signal is given by the complex valued function $f(t)$, then the Fourier transform $F(\omega)$ of $f(t)$ will be given by the expression

$$F(\omega) = \int_{-\infty}^{\infty} f(t)e^{-j\omega t}dt \qquad (1.62)$$

where ω is the radian frequency ($\omega = 2\pi f$). In a naturally symmetric fashion, the time domain function $f(t)$ can be obtained from the frequency domain function $F(\omega)$ by taking the inverse Fourier transform of $F(\omega)$:

$$f(t) = \frac{1}{2\pi} \int_{-\infty}^{\infty} F(\omega)e^{j\omega t}d\omega \qquad (1.63)$$

Some of the notable characteristics that may be obtained from the time domain function include wave period, pulsewidth, time-amplitude target return signature, pulse repetition interval, and dwell time. Notable characteristics that may be obtained from the frequency domain function include wave frequency, signal bandwidth, frequency-amplitude target return signature, pulse repetition frequency, and spectral line bandwidth. For example, jet engine modulation of the radar backscatter and the double frequency backscatter from treaded vehicles can be used for target classification and identification. In addition, the spectral extent of natural clutter return must be known in order to optimally design the high-pass signal filter cutoff frequency for the detection of slowly moving targets in a clutter environment. Figure 1.15 shows the time domain and frequency domain characteristics of an infinite pulse train signal at a single radar center frequency.

The clutter frequency spectrum is often described by the *cutoff frequency*, that frequency f_c where the clutter amplitude is reduced by 3 dB from its maximum value. The cutoff frequency is analogous to the bandwidth of the clutter backscatter signal, as shown in Figure 1.16.

The radar backscatter signal is also commonly characterized by its power density spectrum or its autocorrelation function, which are intimately related to each other. The power density spectrum $P(f)$ of the time signal $X(t)$ over a time period T is defined by the expression

$$P(f) = \lim_{T \to \infty} \frac{1}{2T} \left| \int_{-T}^{T} X(t) \, e^{-2j\pi ft}dt \right|^2 \qquad (1.64)$$

while the autocorrelation function $R(\tau)$ of a stationary (i.e., temporally homogeneous) function $X(t)$ is defined by the expression

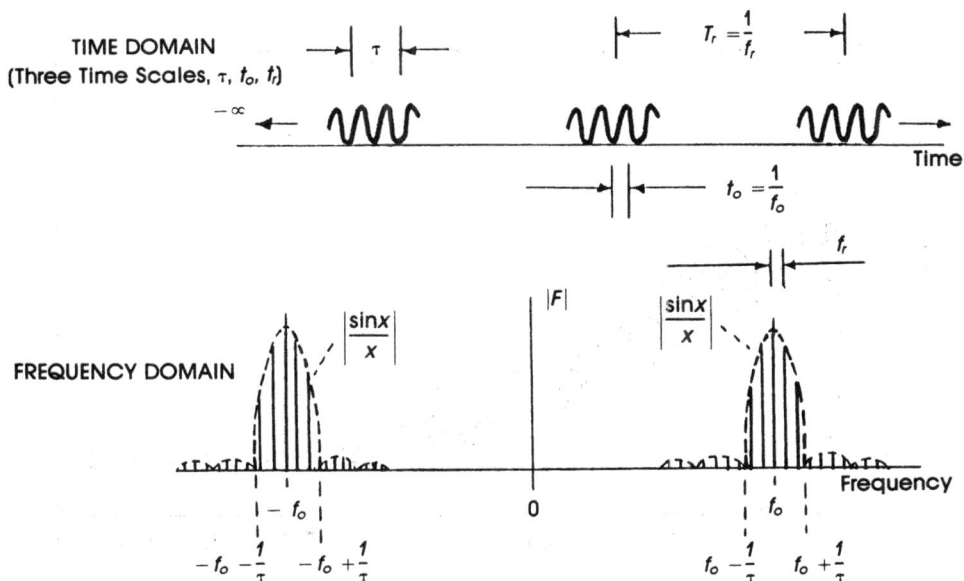

Figure 1.15 Infinite pulse train signal in time (upper curve) and frequency (lower curve) domains. (From Holm [13], © 1987 by Artech House, Inc.)

$$R(\tau) = \lim_{T \to \infty} \frac{1}{2T} \int_{-T}^{T} X(t) \times X(t + \tau)dt \qquad (1.65)$$

The Fourier transform is the power density spectrum of the autocorrelation function (Long [15]):

$$R(\tau) = \int_{\infty}^{\infty} P(f)\, e^{2j\pi ft}df \qquad (1.66)$$

and the total power contained in the spectrum is equal to $R(\tau)$ evaluated at $\tau = 0$:

$$R(0) = \int_{-\infty}^{\infty} P(f)df = 2 \int_{0}^{\infty} P(f)df \qquad (1.67)$$

Figure 1.17 shows the measured, normalized autocorrelation functions for tree backscatter returns at both 6.3 and 35 GHz.

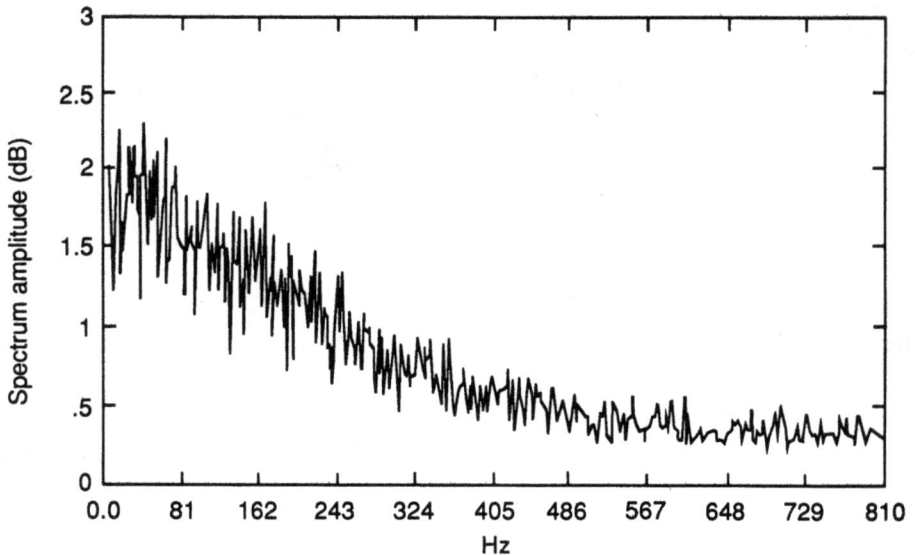

Figure 1.16 Rain spectrum at 70 GHz. The cutoff frequency f_c is approximately 120 Hz. (From Currie, Dyer, and Hayes [14].)

Autocorrelation functions are generally used to compute the decorrelation time of a radar backscattered signal, which is the time for the signal to significantly change its value, because of the changes in the relative phase among the various illuminated scatterers. If the clutter is sampled for a shorter time than this decorrelation time, the new sample provides no new statistical information about the clutter cell and does not improve the detectability of a target in clutter.

Decorrelating the clutter return means that the vector sum clutter backscatter signal is made to vary through all possible phase values. If the surface is made up of several equal-sized scatterers, a Rayleigh distribution is the result, which has a minimum value of 0 and a maximum value of the sum of all the scatterers within the radar beam. The return varies because of changes in phase between the various scatterers along the line of sight to the radar, which can result from either moving the scatterers or changing the signal wavelength. It has been observed that there are often a few dominant scatterers, such as rocks and tree trunks, so that the minimum value is non-zero.

Figure 1.18 presents many scatterers of equal size on a surface, some of which are illuminated by the radar beam, as shown. If just the scatterers along lines of

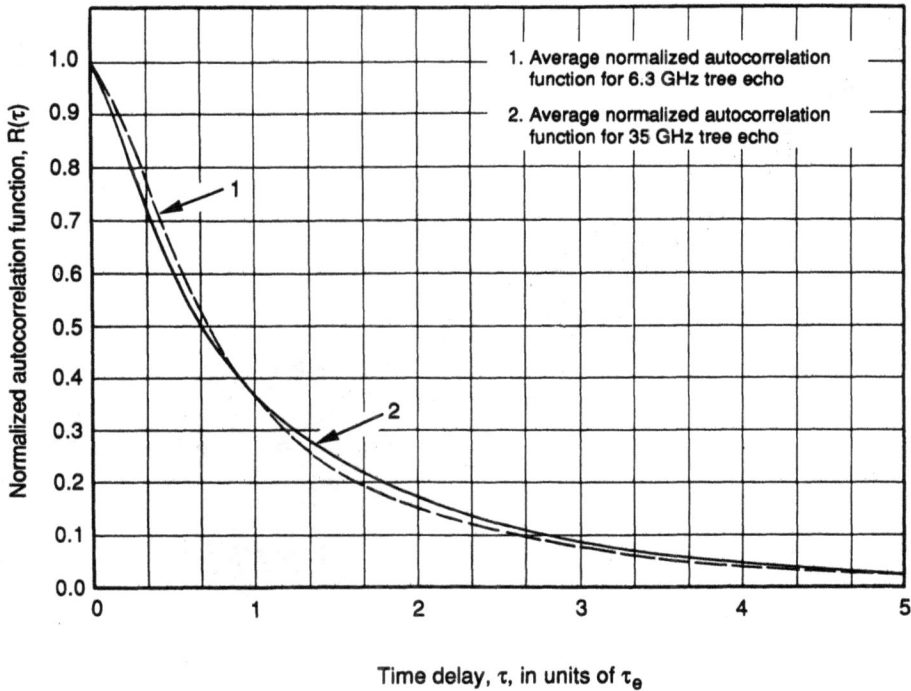

Figure 1.17 Tree backscatter autocorrelation functions at 6.3 and 35 GHz as a function of τ_e, the time for the correlation signals to decay to $1/e$ of their initial values. (From Long, © 1984 by Artech House, Inc. [16].)

equal range to the radar are considered, then their vector sum can be replaced by a single scatterer located at the center of the radar beam, as shown in Figure 1.19.

In order to decorrelate the clutter between radar transmissions, the resulting signal from each of the equivalent scatterers must change a significant amount. As illustrated in Figure 1.20, one way to achieve this aim is to require that the phase between pairs of scatterers within the beam change by the maximum amount. This can be accomplished by changing the signal phase by π radians, which is done by changing the radar frequency between the scatterer pairs.

The two scatterers located at the edges of the illuminated range cell are a distance $c\tau/2$ apart in meters, where c is the speed of light and τ is the pulse length. In order to decorrelate the return on successive interpulse periods, the phase between pairs must change by π radians on successive transmitted pulses by changing the

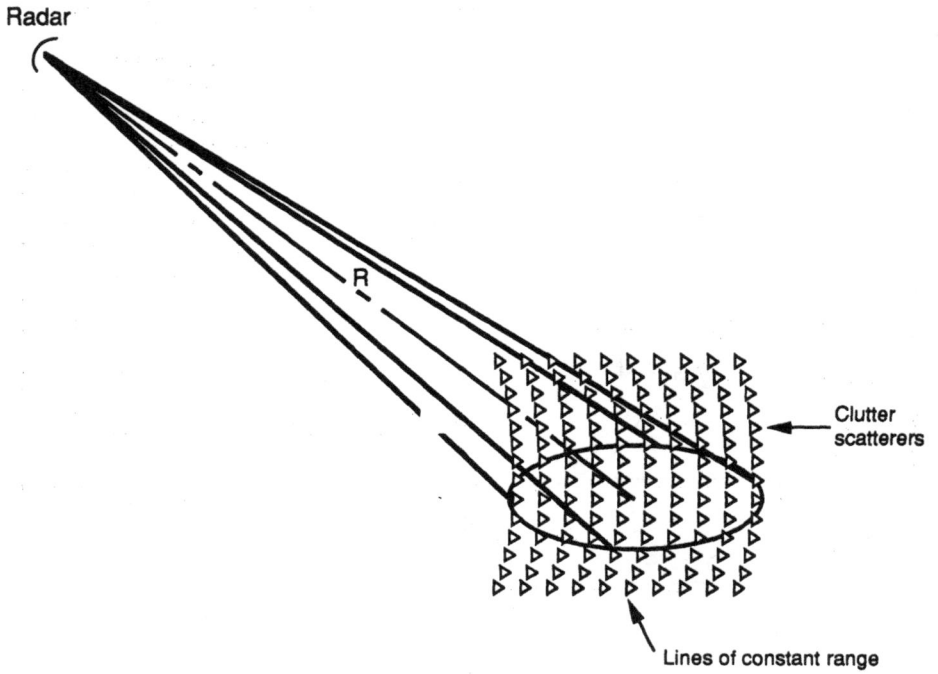

Figure 1.18 Geometry of clutter scatterers.

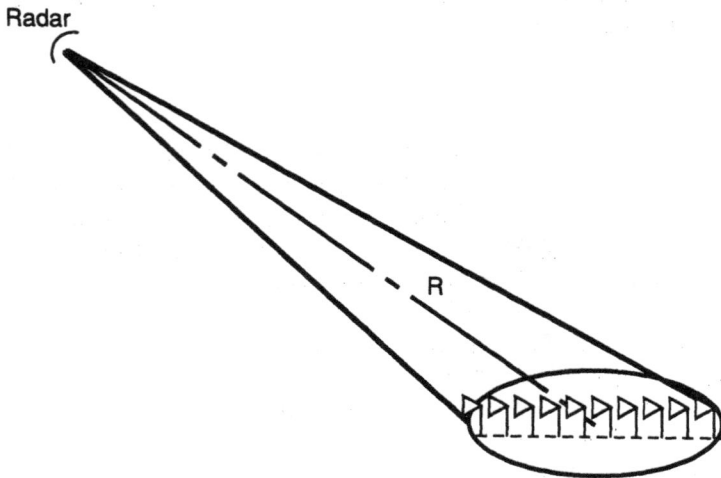

Figure 1.19 Equivalent scatterers located at the center of the radar beam.

frequency. The phase difference $\delta\phi_i$ between two scatterers separated by a distance d_i is given by

$$\delta\phi_i = \frac{4\pi d_i}{\lambda} \tag{1.68}$$

Thus, we can change ϕ_i by changing λ. If we let $\delta\phi_i = \pi$, by changing λ from λ_1 to λ_2, then

$$\delta\phi_1 - \delta\phi_2 = \frac{4\pi d_i}{\lambda_2} - \frac{4\pi d_i}{\lambda_1} = \pi \tag{1.69}$$

and

$$\left[\frac{d_i}{\lambda_2} - \frac{d_i}{\lambda_1}\right] = \frac{1}{4} \tag{1.70}$$

Substituting c/f_i for λ_i in Equation (1.70) gives the expression

$$[f_2 - f_1] = \delta f = \frac{c}{4d_i} \tag{1.71}$$

Two scatterers at the range extent of the beam footprint are separated by a distance d_i equal to the pulse length, $d_i = c\tau/2$. For this case, δf is given by the expression

$$\delta f = \frac{c}{4(c\tau/2)} = \frac{2}{4\tau} = \frac{1}{2\tau} \tag{1.72}$$

Therefore, in order to cause a phase shift of π between the scatterer at the beginning of the range cell and that at the end, the frequency must change by $1/2\tau$. Likewise, it can be shown that a frequency shift of $1/\tau$ is required to cause the phase between the second pair of scatterers indicated in Figure 1.20 to change by π. Table 1.1 summarizes the frequency shifts required to cause the phase between separate pairs of scatterers to change by π.

Thus, frequency shifts of at least $2^{n-2}/\tau$ can decorrelate clutter. For example, for a 100-ns pulse, minimum frequency steps of 5 MHz are required to decorrelate clutter and to obtain 10 independent samples; a total of 50 MHz total stepped band-

Phase between pairs of scatters

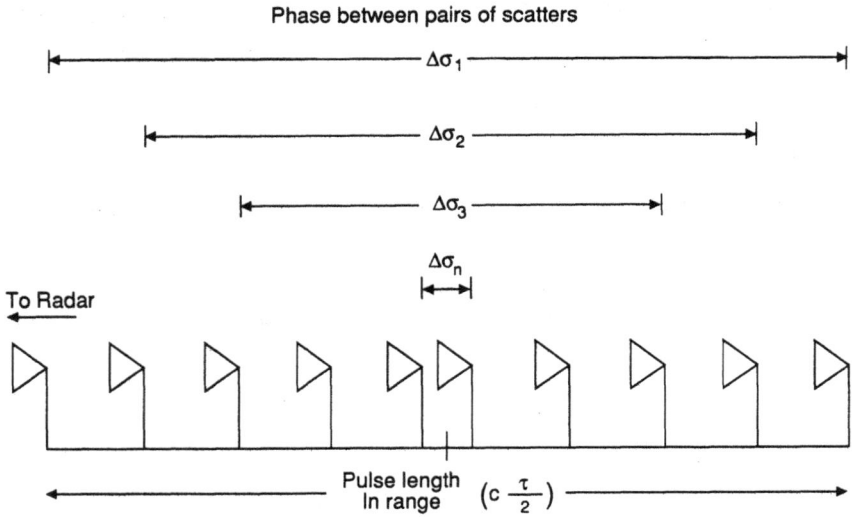

Figure 1.20 Phase between pairs of equivalent scatterers within a pulse length.

Table 1.1
Frequency Shift Required to Change the Phase by π rad for Each Scatterer Pair in Figure 1.20

Scatterer Pair	Frequency Shift Required
$\delta\phi_1$	$1/2\tau$
$\delta\phi_2$	$1/\tau$
$\delta\phi_3$	$2/\tau$
$\delta\phi_n$	$2^{n-2}/\tau$

width would be required. The equally spaced scatterer model assumed above is often valid as illustrated by Figure 1.21, which depicts actual high-range resolution data of the backscatter return from snow-covered ground. The figure presents successive range profiles for four azimuth positions of the beam on the ground, illustrating the many scatterers in range that can be averaged by frequency agility. The range resolution for these data is approximately 0.3 m.

Depression angle = 25 degrees
Azimuth angle = +11 degrees

Depression angle = 25 degrees
Azimuth angle = +12 degrees

Depression angle = 27 degrees
Azimuth angle = +12 degrees

Depression angle = 27 degrees
Azimuth angle = +11 degrees

```
15    12    9    6    3    0
```
Relative range (m)

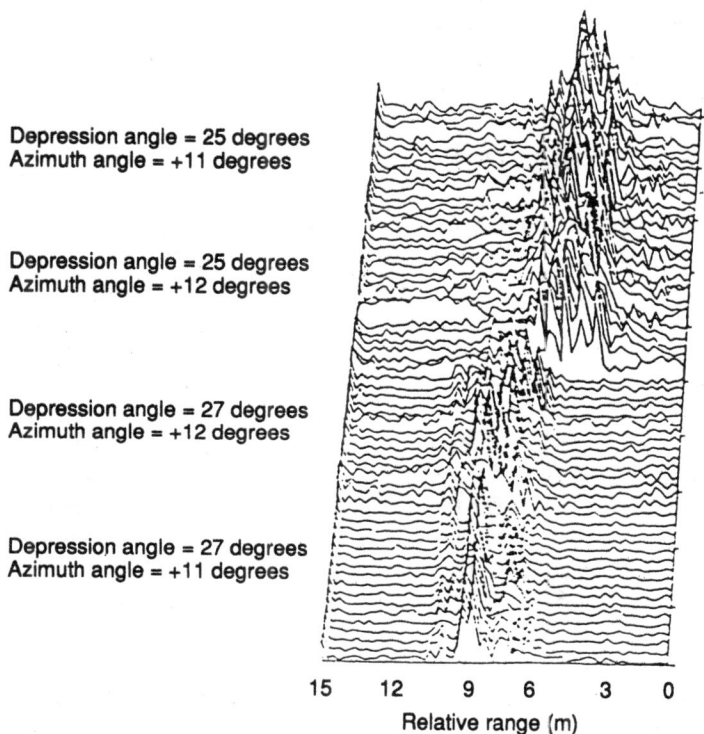

Figure 1.21 Range profile of radar returns at 35 GHz from smooth, snow-covered ground at a 12° depression angle. Each set of 16 range profiles represents the radar return for 20 steps (one beamwidth) in azimuth. (From Currie et al., © 1988 by IEEE [17].)

1.5.3 Spatial Properties

The way that clutter signals vary as the radar beam moves in space can be described in the same manner as the temporal properties (i.e., with frequency spectra and correlation functions). Typically, spatial frequency is not used for radar detection, but spatial correlation is often used to define the number of independent looks at the surface clutter as the radar beam scans across the surface. The number of independent looks is simply the total distance scanned on the ground divided by the spatial decorrelation distance.

From a purely physical point of view, this form of signal decorrelation is the result of the different physical makeup of the totality of scatterers within the clutter resolution cell. As the radar beam scans the clutter surface, some scatterers will no longer be illuminated, while others will become illuminated. The signal received at the radar antenna is the vector sum of the backscattered signals from each of the individual scatterers and will, therefore, change for the same mathematical reason that it changed because of the actual movement of the scatterers themselves in time.

1.6 HOW TO USE THIS BOOK

This book is intended to serve as a primer for the novice on detection in clutter and as a reference for the practicing engineer. Chapter 1 has included introductory material intended primarily for the novice. Chapter 2 summarizes current theory related to clutter properties, with emphasis on millimeter-wave clutter. Chapter 2 should be useful in explaining the physical reasons for various clutter characteristics. Formulas and equations are provided that are useful for modeling target detection in clutter. Chapter 3 contains a summary of available millimeter-wave data on clutter reflectivity and attenuation and provides models available in the literature that are useful for the computer simulation of target detection in a clutter environment. Chapter 4 summarizes the methodology for the detection of targets in clutter and provides for two approaches: statistical detection analysis and computer simulation. Several examples are provided in order to explain the methodologies.

While much of the above material appears in the literature, the authors are unaware of any one competing work that brings all of these concepts together in a single reference text. In particular, standard detection theory cannot be successfully used to solve detection problems involving clutter, so alternative solutions must be developed, as outlined in Chapter 4. The recommended recipe for using this book to solve target-in-clutter detection problems is to follow the procedures outlined in Chapter 4, while drawing on Chapters 2 and 3 for models and data to support the analysis. Additional references helpful in solving detection problems are listed at the end of Chapter 4. Good luck!

REFERENCES

[1]   H.A. Corriher, Jr., et al., "Elements of Radar Clutter," Chapter XVII, *Principles of Modern Radar,* Georgia Institute of Technology Short Course Notes, Atlanta, Georgia, 1972.

[2]   E.F. Knott, "Far Field RCS Measurement Ranges," *Techniques of Radar Reflectivity Measurement,* N.C. Currie, ed., Artech House, Inc., Norwood, Massachusetts, 1984, pp. 250–251.

[3]   M.W. Long, *Radar Reflectivity of Land and Sea,* 2nd ed., Artech House, Inc., Norwood, Massachusetts, 1983, pp. 65–68.

[4]   R.N. Trebits, "Radar Cross Section," *Techniques of Radar Reflectivity Measurement,* N.C. Currie, ed., Artech House, Norwood, Massachusetts, 1984, p. 50.

[5]   R.N. Trebits, "Radar Cross Section," *Techniques of Radar Reflectivity Measurement,* N.C. Currie, ed., Artech House, Norwood, Massachusetts, 1984, p. 48, after F.E. Nathanson, *Radar Design Principles,* McGraw-Hill Book Company, Inc., New York, 1969, p. 65.

[6]   R.N. Trebits, "Radar Cross Section," *Techniques of Radar Reflectivity Measurement,* N.C. Currie, ed., Artech House, Norwood, Massachusetts, 1984, p. 49.

[7] W.K. Rivers, "Low Angle Sea Return at mm Wavelength," Final Report on Contract N62267–70-C-0489, Georgia Institute of Technology, Atlanta, Georgia, 1970.

[8] J. Aitchison and J.A.C. Brown, *The Log-Normal Distribution*, Cambridge University Press, London, 1969.

[9] M.W. Long, "Statistical Properties of Data," *Techniques of Radar Reflectivity Measurement*, Artech House, Inc., Norwood, Massachusetts, 1984, pp. 433–460.

[10] D.E. Kerr, *Propagation of Short Radio Waves*, McGraw-Hill Book Company, Inc., New York. 1951.

[11] J. Aitchison and J.A.C. Brown, *The Log-Normal Distribution*, Cambridge University Press, London, 1969.

[12] R.D. Hayes, personal communications, 1991.

[13] W.A. Holm, "MMW Radar Signal Processing Techniques," Chapter 6 in *Principles and Applications of Millimeter-Wave Radar*, N.C. Currie and C.E. Brown, eds., Artech House, Inc., Norwood, Massachusetts, p. 246.

[14] N.C. Currie, F.B. Dyer, and R.D. Hayes, "Analysis of Radar Rain Return at Frequencies of 9.375, 35, 70, and 95 GHz," Technical Report No. 2 on Contract DAAA 25-73-C-0256, Georgia Institute of Technology, Atlanta, Georgia, 1975, p. 79.

[15] M.W. Long, *Radar Reflectivity of Land and Sea*, 2nd ed., Artech House, Inc., Norwood, Massachusetts, 1983, p. 144.

[16] H.D. Ivey, M.W. Long, and V.R. Widerquist, "Polarization Properties of Echoes From Vehicles and Trees," *Record of the Second Annual Radar Symposium*, University of Michigan, 1956.

[17] N.C. Currie et al., "Millimeter Wave Measurements and Analysis of Snow-Covered Ground," *IEEE Trans. Remote Sensing and Geoscience*, Vol. 26, No. 2, May 1988, pp. 307–318.

Chapter 2
Backscatter and Attenuation Theory

2.1 INTRODUCTION

This chapter covers the fundamental theories that apply to radar clutter: the electromagnetic scattering from both single and distributed objects and the loss of signal energy through a medium filled with natural or manmade materials. We will start with the description of the radar backscatter from the most mathematically tractable object, the conducting sphere, and then move on to ensembles of scatterers made up of hydrometeors: fog, rain, and snow. Finally, we will describe the attenuation of signal energy in clear air; in fog, rain, and snow; and in natural and manmade obscurant materials.

2.2 SCATTERING FROM SPHERES

The perfectly conducting sphere has total symmetry and is one of the few mathematically tractable targets for which the radar cross section (RCS) may be calculated in closed form. This physical problem can be formulated into a differential equation of the form

$$z^2 \frac{d^2w}{dz^2} + z \frac{dw}{dz} + (z^2 - n^2)w = 0 \tag{2.1}$$

where the function w depends on the variable z.

The exact solution for the RCS of a sphere of radius a, usually referred to as the Mie series, is given by

$$\sigma = \frac{\lambda^2}{\pi} \left| \sum_{n=1}^{\infty} (-1)^n (n + 1/2)(b_n - a_n) \right|^2 \tag{2.2}$$

---

where λ is the radar wavelength, and the coefficients a_n and b_n are given by

$$a_n = \frac{J_n(ka)}{H_n^{(1)}(ka)} \tag{2.3}$$

$$b_n = \frac{kaJ_{n-1}(ka) - nJ_n(ka)}{kaH_{n-1}^{(1)}(ka) - nH_n^{(1)}(ka)} \tag{2.4}$$

In Equations (2.3) and (2.4), wavenumber notation k is used, where

$$k = \frac{2\pi}{\lambda} \tag{2.5}$$

$J_n(ka)$, $Y_n(ka)$, and $H_n^{(1)}(ka)$ are the spherical Bessel functions of the first, second, and third kind, respectively, of order n and argument ka, using standard mathematical notation [1, 2]. The functions $Y_n(ka)$, also called Weber's functions, and the functions $H_n^{(1)}(ka)$, also called Hankel functions, are related to $J_n(ka)$ using the relationships

$$H_n^{(1)}(ka) = J_n(ka) + iY_n(ka) \tag{2.6}$$

and

$$Y_n(ka) = \frac{J_n(ka)\cos(n\pi) - J_{-n}(ka)}{\sin(n\pi)} \tag{2.7}$$

Figure 2.1 depicts the RCS of a perfectly conducting sphere normalized by its geometric cross section πa^2 as a function of the parameter ka. From Equation (2.5), we see that the parameter ka is just $2\pi a/\lambda$, the sphere's circumference in wavelengths.

It is readily apparent from Figure 2.1 that there are three very distinct wavelength regions for this conducting sphere. We see that for low frequencies, where the radar wavelength is much greater than the sphere's circumference (the Rayleigh region, where $2\pi a/\lambda \ll 1$), the sphere's RCS varies proportionately to $(ka)^4$. Likewise, where the radar wavelength is much smaller than the sphere's circumference (the optical region, where $2\pi a/\lambda \gg 10$), the sphere's RCS is essentially wavelength-independent and asymptotically approaches a value of πa^2, its geometric cross section. In between these two frequency regions (the Mie region, where $1 \ll 2\pi a/\lambda \ll 10$), the sphere's RCS oscillates about its geometric cross section value. This oscillatory character can be explained in a physical manner as the vector summation of the specular radar reflection from the "front" of the sphere with a radar wave that

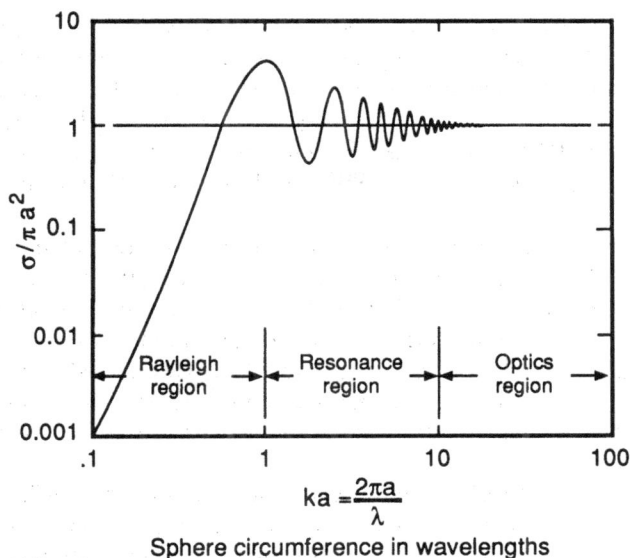

$$ka = \frac{2\pi a}{\lambda}$$

Sphere circumference in wavelengths

Figure 2.1 The normalized RCS of a perfectly conducting sphere as a function of circumference. (From Knott, Shaeffer, and Tuley, ©1985 by Artech House, Inc.)

"creeps" around the shadowed side of the sphere. Constructive or destructive interference results when the two signals combine in space along the line of sight to the radar receiver.

Clearly, then, objects with significant dimensions between 1 and 10 mm will have RCS values in the millimeter-wave part of the electromagnetic spectrum and will be different from that at lower, microwave frequencies. Other than differences in apparent "roughness" and dielectric values, objects that were "small" compared to a wavelength at microwave frequencies may be "large" at millimeter wave frequencies. For example, rain backscatter can be treated as a Rayleigh scattering phenomenon at microwave frequencies, since they have diameters up to several millimeters. However, rain backscatter at millimeter wavelengths is clearly within the Mie region, and a different mathematical strategy must be employed to calculate RCS values.

2.3 SCATTERING FROM NONSPHERICAL OBJECTS

The frequency dependence of many nonspherical object's RCS can be generalized from Figure 2.1. In this case, we take the "size" L, or significant dimension, as the equivalent to the parameter a that was used to characterize the RCS for a sphere of radius a. The low-frequency region ($kL \ll 1$) is called the *Rayleigh region*, after Lord Rayleigh, and is characterized by the same fourth-power dependence of a sphere's RCS on frequency. The mid-frequency region ($1 \ll kL \ll 10$), characterized by an oscillatory dependence of RCS on frequency, is referred to as the Mie, or *resonance region*. Finally, the RCS in the high-frequency region ($kL \gg 10$) does not vary significantly from its geometric cross section with frequency and is called the *optics region*.

In general, the RCS of an object any more complicated than a perfectly conducting sphere will be a function of the object's (1) dielectric properties, (2) roughness, and (3) shape. Shape-related factors will include specular reflections from "flat, smooth" surfaces, diffuse scattering from "rough" surfaces, diffraction scattering from edges and corners, and "creeping waves." Exact, closed-form computations are typically not mathematically feasible, but lend themselves to a host of predictive techniques, including physical optics, geometric optics, and geometric theory of edge diffraction (GTD).

Most equations for RCS in the literature assume that the pertinent dimensions of the target object are large compared to the radar wavelength (*i.e.*, the target's optics RCS). The RCSs of some of the more commonly reported shapes are included in Table 2.1 [3].

The strength of the various reflectivity mechanisms can be ordered by rank into a hierarchy of scattering shapes, and are listed in Table 2.2. Corner reflectors are

Table 2.1
Radar Cross Section of Simple Geometric Objects
(Adapted from Trebits, Ref.[3], ©1984 by Artech House)

Shape	Radar Cross Section Formulation	Conditions
Sphere	πa^2	$2\pi a/\lambda > 10$ a = radius of sphere
Flat Circular Plate	$\dfrac{4\pi^3 a^4}{\lambda^2} \left[2\dfrac{J_1(u)}{u} \right]^2 \cos^2\phi$	$u = (4a/\lambda) \sin\phi$ ϕ = angle to normal a = radius of plate at normal incidence
Flat Plate	$4\pi(S/\lambda)^2$	S = surface area
Cylinder	$\dfrac{2\pi a l^2}{\lambda} \left[\dfrac{\sin N}{N} \right]^2 \cos\phi$	a = radius ϕ = angle to normal $N = (2\pi l/\lambda) \sin\phi$
Cone	$\pi a^2 \tan^2\alpha$	α = half-angle of cone a = base radius, normal incidence

Table 2.2
Hierarchy of Scattering Shapes, in Descending Strength
(Adapted from Knot, Shaeffer, and Tuley, ©1985 by Artech House, Inc. [4])

Geometry	Type	Freq. Dep.	Size Dep.	Formula	Remarks
	Square trihedral corner retroreflector	F^2	L^4	Maximum $\sigma = \dfrac{12\pi a^4}{\lambda^2}$	Strongest return; high RCS due to triple reflection
	Right dihedral corner reflector	F^2	L^4	Maximum $\sigma = \dfrac{8\pi a^2 b^2}{\lambda^2}$	Second strongest return; high RCS due to double reflection, tapers off gradually from the maximum with changing θ and sharply with changing σ
	Flat plate	F^2	L^4	Maximum $\sigma = \dfrac{4\pi a^2 b^2}{\lambda^2}$ Normal incidence	Third strongest return; high RCS due to direct reflection, drops off sharply as incidence changes from normal

Table 2.2 (cont'd)

Geometry	Type	Freq. Dep.	Size Dep.	Formula	Remarks
	Cylinder	F^1	L^3	Maximum $\sigma = \dfrac{2\pi ab^2}{\lambda}$ Normal incidence	Prevalent cause of strong, broad RCS over varying aspect (θ), drops off sharply as azimuth (ϕ) changes from normal; can combine with flat plate to form dihedral corner reflector
	Sphere	F^0	L^2	Maximum $\sigma = \pi a^2$ Normal incidence	Prevalent cause of strong, broad RCS peaks other than those due to large openings in target body; energy defocused in two directions
	Straight edge normal incidence	F^0	L^2	$f(\theta, \phi)L^2$ θ—aspect	Limiting cause of 2-dimensional curved plate mechanism as radius

Mechanism			Parameters / Formula	Description
			θ_{int}—interior dihedral angle between faces meeting at edge	shrinks to 0, prevalent cause of strong, narrow RCS peaks form supersonic aircraft
Curved edge normal incidence	F^{-1}	L^1	$f(\theta, \theta_{int})a\lambda/2$ $a \geq \lambda$	Limiting cause of 3-dimensional curved plate mechanism as principal radius shrinks to 0; the function f is the same as in mechanism 3
Apex	F^{-2}	L^0	$\lambda^2 g(a, P, \theta, \phi)$ a,β—interior angles of tip θ,ϕ—aspect angles	Limiting cause of previous mechanism as a shrinks to 0, for $a = b$, the tip is that of a cone, for $a = 0$, the tip is the corner of a thin sheet or fin
Discontinuity of curvature along a straight line, normal incidence	F^2	L^0	$\dfrac{\lambda^2}{64\pi^3}\left(\dfrac{L}{a}\right)^2\left(1+\left(\dfrac{dy}{dx}\right)^2\right)^{-3/2}$ $(1/a)$—jump in reciprocal of dy/dx—slope of surface w.r.t. incident ray	Strongest of an infinite sequence of discontinuities; very weak mechanism, which together with 6, shares dominance of nose-on RCS of cone sphere

Table 2.2 (cont'd)

Geometry	Type	Freq. Dep.	Size Dep.	Formula	Remarks
	Discontinuity of curvature of a curved edge	F^{-3}	L^{-1}	$f(\theta,\phi)\dfrac{\lambda^3 b}{a^2}\left\{1+\left(\dfrac{dy}{dx}\right)^2\right\}^{-3/2}$ $f(\theta,\phi)$—function of aspect b—radius of edge $>\lambda$	Important mechanism for traveling wave backscatter where RCS of discontinuity is augmented by gain of traveling wave structure, dependences are based on dimensional considerations
	Discontinuity of curvature along an edge	F^{-4}	L^{-2}	$g(\theta,\phi)\lambda^2\dfrac{1}{a}\left\{1+\left(\dfrac{dy}{dx}\right)^2\right\}^{-3/2}$ $g(\theta,\phi)$—function of aspect	Important mechanism for traveling wave backscatter where RCS of discontinuity is augmented by gain of traveling wave structure, dependences are based on dimensional considerations

the first set (and the largest scatterers) in this hierarchy and are often used for reference targets and RCS measurement calibration. Rays incident on a corner reflector are reflected back along the direction from which they came. The RCS of a corner reflector varies with the square of the radar frequency. The next set of scatterers includes the flat plate, the cylinder, and the sphere, all of which demonstrate a specular scattering mechanism (e.g., an area, line, or spot that reflects "brightly"). The next smaller reflective scatterers include the straight edge, the curved edge, and the tip. Finally, the smallest of the scattering sets includes discontinuities in surface curvature, where no portion of the surface is oriented normally to the radar line of sight [4].

2.4 SCATTERING FROM DISTRIBUTED OBJECTS

For the most part, radar clutter consists not so much of large RCS, isolated point objects, but of a large number of individual, small RCS scatterers. The surface of the ground, the sea surface, and a volume filled with falling rain drops all represent ensembles of scattering centers, which together create the radar backscatter signal competing for detection with the backscattered signal from the desired target. Figure 2.2 shows a two-dimensional ensemble of scatterers illuminated by a radar.

The backscattered vector electromagnetic signal from each scatterer will have an amplitude and phase $\mathbf{E}_i$ corresponding to a particular polarization, where $\mathbf{E}_i$ is given by

$$\mathbf{E}_i = \mathbf{A}_i e^{j\Phi_i} \tag{2.8}$$

Recall from Chapter 1 that the phase of the electromagnetic signal has both temporal and spatial dependencies, in addition to a constant term:

$$\Phi = \omega t - kR + \phi \tag{2.9}$$

where $\omega = 2\pi f$ is the radar radian frequency and $k = 2\pi/\lambda$ is the wavenumber. The temporal component of the phase for each scatterer will change with time. The spatial component will change only if the spatial relationship among the scatterers changes, either because the scatterers themselves are moving or because the radar is changing its aspect with respect to the scatterer ensemble.

The composite electromagnetic signal $\mathbf{E}$ which is received at the radar antenna will be the *vector* sum of the individual contributions, as shown in Figure 2.3, and expressed mathematically in Equation (2.10).

$$\mathbf{E} = \sum_{i=1}^{n} \mathbf{E}_i = \sum_{i=1}^{n} \mathbf{A}_i e^{j\Phi_i} \tag{2.10}$$

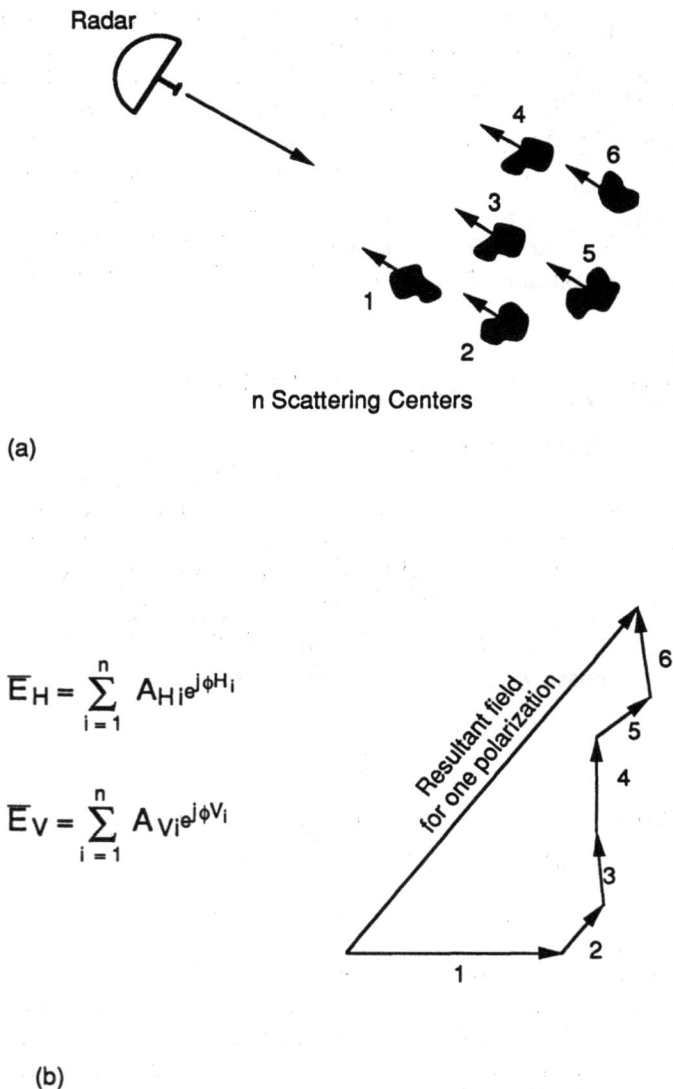

(a)

$$E_H = \sum_{i=1}^{n} A_{Hi} e^{j\phi_{Hi}}$$

$$E_V = \sum_{i=1}^{n} A_{Vi} e^{j\phi_{Vi}}$$

(b)

Figure 2.2 The vector summation of signals reflected by individual scatterers. (a) Vector-phasor summation of individual field contributions. (b) Clutter return from an assembly of scattering centers. (From Corriher et al., ©1980 by Georgia Institute of Technology.)

It should be obvious then that the composite electromagnetic signal **E** can vary considerably with even small changes in scatterer position, radar aspect, or radar frequency. This phase sensitivity is even more apparent at millimeter-wave frequencies than it is at microwave frequencies, since millimeter wavelengths are one-tenth as large.

2.4.1 Scattering From Hydrometeors

The class of meteorological scatterers of interest to radar engineers includes fog, clouds, haze; rain, snow, and hail. Fog, rain, and snow backscatter will be addressed; hail occurs infrequently and will not be covered here.

2.4.1.1 Fog and Haze Backscatter

Fog and haze are meteorological conditions that represent liquid water suspended in the atmosphere. Fog is differentiated from haze purely on the basis of visibility, where fog is defined as a condition limiting visibility to 1 km or less. Visibility is typically defined as the distance at which "large, dark objects can be discerned against the horizon by human observers" [6].

Two fog types are generally identified in the literature: radiation fogs and advection fogs. Radiation fogs form when the ground cools after sunset, and the adjacent air mass cools until the air becomes supersaturated with water. In contrast, advection fogs form when warm, moist air overrides a cooler surface. Figure 2.3 depicts visibility versus liquid water density in the air for both advection and radiation fogs [6].

Of the radar signal energy scattered near the target location, the backscattered energy (i.e., the energy scattered backward, along the radar line of sight) is quite obviously the most relevant to radar performance. The fog backscatter coefficient η can be derived from the Rayleigh scattering theory approximation to the Mie solution for scattering from spheres:

$$\eta = \frac{\pi^5 |K|^2}{\lambda^4} Z \qquad (2.11)$$

where

η = backscatter coefficient, in units of m^2/m^3
λ = the radar wavelength, in mm
$|K|$ = the refractive index factor = $(m^2 - 1)/(m^2 + 2)$

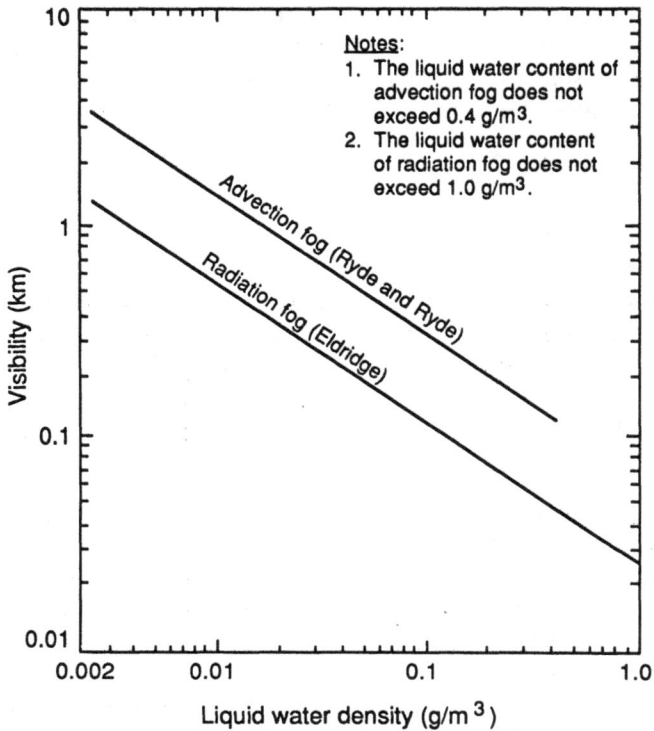

Figure 2.3 Optical visibility in fogs versus liquid water density. (Adapted from Eldridge [7].)

M = the liquid water content per unit volume of fog, in units of g/m^3
m = complex index of refraction of water

Z is usually referred to as the reflectivity factor. It has units of (mm)^6/m^3 and is described by one of the following expressions (© 1964 Academic Press):

$$Z = 0.48 \, M^2 \quad \text{for a radiation fog}$$

$$Z = 8.2 \, M^2 \quad \text{for an advection fog [8]}$$

At millimeter wavelengths, the fog reflectivity is less than a hundredth of that from rain, and will have essentially no effect on short-range radar detection or tracking performance.

2.4.1.2 Rain Backscatter

The rain backscatter coefficient may be empirically described in a format similar to that for attenuation:

$$\eta = AR^B \tag{2.12}$$

where η is the backscatter coefficient (m^2/m^3), R is the rainfall rate in mm/hr, and A and B are empirically derived parameters. The extraordinary drop size dependency on rain backscatter is apparent from the relationship

$$\eta = \frac{\pi^5 |K|^2}{\lambda^4} \sum D^6 \tag{2.13}$$

where D is the drop diameter, $|K|$ is the refractive index factor (a function of the dielectric constant of the rain drops), and λ is the radar wavelength. Thus the backscatter coefficient is strongly influenced by the presence of larger size drops.

Parenthetically, consider the graph of the normalized RCS of a spherical rain drop as a function of both drop size and radar frequency, as shown in Figure 2.4. The RCS data for a perfectly conducting sphere, from Figure 2.1, are included in Figure 2.4, whose data were calculated using the dielectric properties of water at 20°C. Note that the water cross section curves are lower in magnitude than that for the perfectly conducting sphere. Also, a shift in the frequency sensitivity with drop size is observed for the dielectric sphere. Table 2.3 lists expressions for the RCS for a perfectly conducting sphere and a dielectric sphere in both the Rayleigh region and the optical region, where μ and ε are the permittivity and permeability, respectively, of the dielectric material.

An alternative form of Equation (2.10) often found in the literature is

$$\eta = \frac{\pi^5}{\lambda^4} |K|^2 Z (10^{-6}) \tag{2.14}$$

where Z is called the reflectivity factor, and for Rayleigh scattering is given by

$$Z = \sum_i D^6 N(D_i) \Delta D \tag{2.15}$$

where $N(D_i)\Delta D$ is the number of drops of diameter D_i in the increment ΔD. For millimeter-wave applications, where Mie scattering theory is more appropriate, the

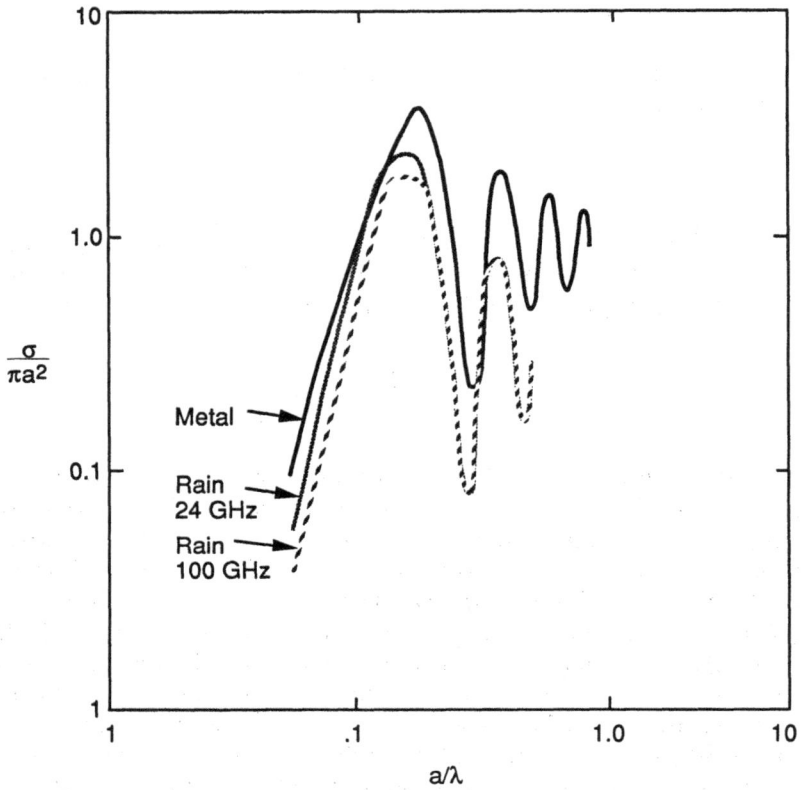

Figure 2.4 Normalized RCS of spheres as functions of ratio of radius to wavelength.

Table 2.3
RCS of a Perfectly Conducting Sphere and a Dielectric Sphere

	Rayleigh Region	*Optical Region*
Conducting Sphere	$9\pi \left(\dfrac{2\pi}{\lambda} \right)^4 a^6$	πa^2
Dielectric Sphere	$4\pi \left(\dfrac{2\pi}{\lambda} \right)^4 \left| \dfrac{\varepsilon - 1}{\varepsilon + 2} \right|^2 a^6$	$\pi \left| \dfrac{\sqrt{\mu\varepsilon} - 1}{\sqrt{\mu\varepsilon} + 2} \right|^2 a^2$

expression for Z is given by

$$Z = \sum_i F_i D^6 N(D_i) \Delta D \tag{2.16}$$

where F_i is the ratio of Mie to Rayleigh scattering for the spherical drop [9].
The general form of Equation (2.16) ultimately becomes

$$Z = A R^B \tag{2.17}$$

which is universally known in the literature as the Z-R relationship. Values for the parameters A and B depend on wavelength, rainfall rate, and on temperature. Representative values for millimeter waves and moderate rainfall rates, extrapolated from microwave values, are approximately 200 for A and 1.1 for B.

In general, Mie scattering theory is appropriate for calculating RCS parameters for particles the size of rain drops. The backscatter coefficient η will then be given by

$$\eta = \int_{D_{min}}^{D_{max}} \sigma(D) n(D) dD \tag{2.18}$$

where

$\sigma(D)$ = backscatter cross section of particles with diameter D
$n(D)dD$ = the number of particles having diameters between D and $D + dD$ per unit volume

Thus, $n(D)$ is the distribution of rain drop sizes within the appropriate radar resolution cell. Two widely cited rain drop distribution models are those of Marshall and Palmer and of Laws and Parsons. The Marshall-Palmer distribution proposes an exponential relationship between the parameter $n(D)$ and the rain drop diameter D [10]:

$$n(D)da = N_0 e^{-\Lambda D} dD \tag{2.19}$$

where

$N_0 = 0.08$ cm^{-4}
$\Lambda = 41 R^{-0.21}$ cm^{-1}
R = rainfall rate, in mm/hr

Other researchers have modeled the rain drop size distribution in similar fashion, with expressions for N_0 that may include rainfall rate dependance and expressions for Λ similar to that in Equation (2.19).

The Laws and Parsons distribution, in contrast, proposes a relationship,

$$n(a)da = \frac{10^3 Rm(a)da}{4.8\pi a^3 v(a)} \qquad (2.20)$$

where

R = rainfall rate, in mm/hr
$v(a)$ = rain drop fall velocity, in m/s
a = rain drop radius
$m(a)$ = percentage of the total resolution volume reaching the ground, contributed by drops of different sizes [11]

The effect that the drop size distribution can have on rain backscatter is especially notable during the passage of a summertime rainstorm. The nature of such a rainstorm is that larger drops dominate the backscattered energy early in the history of the storm's passage, attended by very heavy rainfall rates, often exceeding 100 mm/hr. Later on in the storm's passing, the rainfall rate decreases, and the drop size distribution shifts downward toward smaller sizes. Toward the end of a storm, the drop sizes are quite small as the rainfall rate becomes simply a drizzle. The history of radar backscatter matches the rain drop size distribution history, great at the leading edge of the storm, then steadily decreasing as the storm passes over the observation site.

Figure 2.5 depicts the measured average rainfall backscatter coefficient η at X-band (9.375 GHz) and at 35, 70, and 95 GHz over a wide range of rain rates up to 100 mm/hr [12].

2.4.1.3 Snow Backscatter

Backscatter data recorded from falling snow at millimeter wavelengths is considerably less plentiful than that from rain. However, several general observations have been made. Snow reflectivity is strongly dependent on snow concentration (mass of snow per unit volume) up to 0.2 gm/m^3 and is less dependent for concentration values above this rate. In addition, snow reflectivity values increase with increasing radar frequency (i.e., 140 GHz backscatter is higher than that at 95 GHz), but there does not appear to be much difference between reflectivity at 140 and 225 GHz.

It has also been noted that reflectivity was higher when snow was mixed with rain than snow alone. Rain backscatter was normally higher than snow, given equal

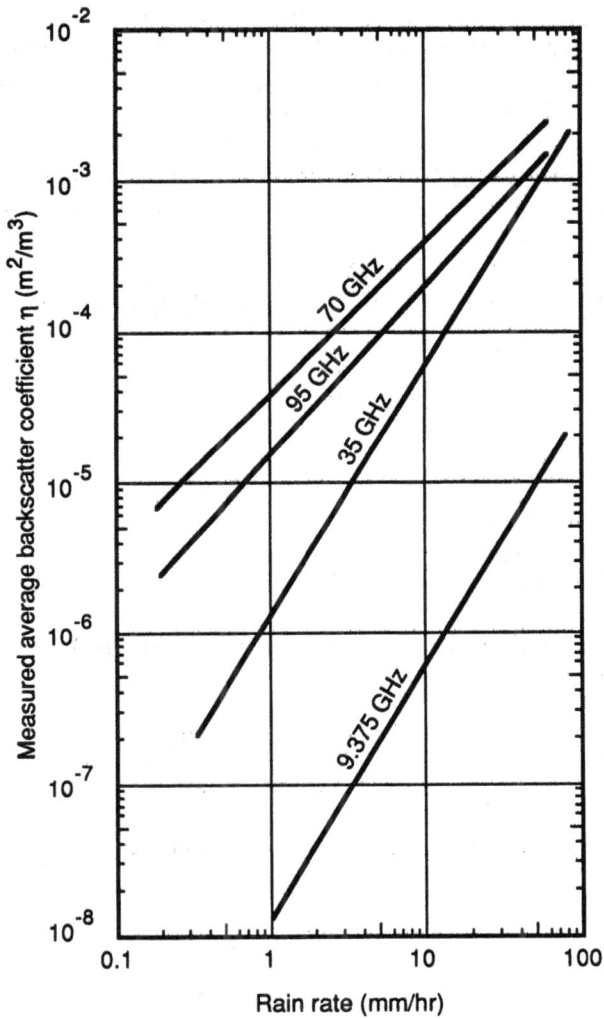

Figure 2.5 Measured backscatter coefficient of rain versus rain rate and radar frequency. (Adapted from Currie, Dyer, and Hayes [12].)

mass concentrations, for two reasons: (1) the value of $|K|^2$ for ice is 0.197, and (2) snowfall rates are usually less than rainfall rates. Radar reflectivity from the so-called *bright band* (the altitude in a rain storm where the temperature is just below 0°C) can be significantly higher than altitudes above or below this level. This is due to the melting of ice or snow particles below the freezing altitude level. These melting particles have a higher reflectivity than the frozen particles above this level or the liquid drops below this level because the drops are smaller and the quantity of water per unit volume is smaller.

The width of the snow fall velocity spectrum is narrow compared to that of rain, approximately 1 m/s [13]. Gunn and Marshall modeled the snow size distribution function $n(D)$ [14]:

$$n(D)da = N_0 e^{-\Lambda D} dD \qquad (2.21)$$

where

$N_0 = 0.038\ R^{-0.87}\ \mathrm{cm}^{-4}$
$\Lambda = 25.5\ R^{-0.48}\ \mathrm{cm}^{-1}$
R = equivalent rainfall rate, in mm/hr

Sekhon and Srivastava [15] modeled the snow size distribution function $n(D)$ with the parameters:

$$N_0 = 0.025\ R^{-0.94}\ \mathrm{cm}^{-4}$$

$$\Lambda = 22.9\ R^{-0.45}\ \mathrm{cm}^{-1}$$

Several published Z-R relationships for snowfall, at microwave frequencies, include

$$Z = 1780\ R^{2.21}\ [15] \qquad (2.22)$$

$$Z = 1000\ R^{1.6}\ [16] \qquad (2.23)$$

$$Z = 2000\ R^2\ [14] \qquad (2.24)$$

where Z has units of $(\mathrm{mm})^6/\mathrm{m}^3$ and R has units of mm/hr (equivalent rainfall rate).

2.4.2 Particulate, Smoke, and Aerosol Scattering

Battlefield obscurants include dust and ejected dirt produced by vehicular traffic and munitions, smoke produced by fires, as well as fog and smoke purposefully deployed to produce sensor obscuration. Wind-blown dust and sand particulates can remain suspended in the air for significant periods of time, while soil ejecta caused by explosions remain airborne for at most several seconds because of their significant mass and the usual effect of gravity. Typical manmade obscurants include [17]:

HC (hexachloroethane)
WP (white phosphorus)
RP (red phosphorus)
Oil fog
HE (high explosive) dust
Vehicular dust
Aluminum flakes
Carbon flakes
Smoke (from vehicles)

Small particulates, like dust, have sizes with dimensions in the range of 1 to 10 μm. Thus, the particulate size is small compared to the radar wavelength, and Rayleigh approximations are appropriate. The backscatter coefficient η for a volume of particulates of diameter D will then be given by the usual expression

$$\eta = \frac{2\pi^5|K|^2}{3\lambda^4} \sum_D D^6 N \, dD \qquad (2.25)$$

where

$K = (m^2 - 1)/(m^2 + 2)$
m = complex index of refraction of the material
N = number of particles in the size interval D and $D + dD$

The equivalent expression to Equation (2.25) for the actual RCS σ of a particle of radius a is given by [18]

$$\sigma = \frac{8}{3} \pi k^4 a^6 |K|^2 \qquad (2.26)$$

where $k = 2\pi/\lambda$.

Limited quantitative information is available in the literature regarding particle size distributions, particularly for sand and dust in sand storms. Three classes of earth particles are defined: silt, sand, and grains of earth. Misers Bluff II test data indicated three size regions also, called small mode, large mode, and ballistic mode, which roughly correspond to the previously named size regions. The small-mode particles are mostly silt and have sizes that are lognormally distributed. The large-mode particles are mostly sand and grains and have sizes that are also lognormally distributed. The ballistic-mode particles are composed mostly of grains of dirt and soil and have sizes that are exponentially distributed. Table 2.4 lists the pertinent statistical and material parameters for these three regions [19], where N is the number of particles per unit volume, M is the particle mass per unit volume, a_g is the particle radius mode value, a_m is the radius mean value, and s is the standard deviation of the distribution of radius values.

Hayes has computed the radar backscatter coefficient η for sand that is typical of sand storms, which is large mode, assuming both quartz and silica composition. At X-band (10 GHz), the value for η is 4.0×10^{-11} m^{-1} for both materials, while at K$_a$-band (35 GHz) η is 6.0×10^{-9} m^{-1} for silica and 7.4×10^{-9} m^{-1} for quartz. Similar calculations for ballistic-mode particles yield X-band η values of 0.5×10^{-7} m^{-1} for both materials and 35 GHz values of 76×10^{-7} m^{-1} for silica and 95×10^{-7} m^{-1} for quartz [18].

The backscatter coefficient η for small particles is seen to be functionally dependent on the complex valued index of refraction, or, equivalently, on the dielectric constant. Thus, η is proportional to $|K|^2$, or

$$\eta \propto |K|^2 = \left| \frac{m^2 - 1}{m^2 + 2} \right|^2 = \left| \frac{\varepsilon - 1}{\varepsilon + 2} \right|^2 \tag{2.27}$$

where ε is the dielectric constant for the particle material.

Scattering efficiencies are quantified in units of the scattering coefficient (m^2/ m^3) per gram of material per cubic meter of air. Mie calculations of the scattering and backscatter efficiencies α_s and α_b at 35 and 95 GHz, for various size particles with refractive indices characteristic of soil particles, are shown in Table 2.5 [20].

Table 2.4
Statistical and Material Parameters for the Particle Regions
(Adapted from Thomas and Cockayne [19])

Parameter	Small Mode Particles	Large Mode Particles	Ballistic Mode Particles
Distribution	Lognormal	Lognormal	Exponential
N/cm^3	200	0.07	—
M (g/m^3)	0.016	0.049	—
a_g (μm)	0.5	22.5	—
a_m (μm)	7.75	73	—
s (μm)	2.6	1.87	—
Time in air	Hours	Minutes	Seconds

Table 2.5
Scattering and Backscatter Efficiencies α_s and α_b, Respectively,
of Soil Particles at 35 and 95 Ghz [51]

	Efficiency	Radius a = 50 μm	Radius a = 300 μm
35 GHz	α_s	7.2×10^{-9}	1.6×10^{-6}
	α_b	1.1×10^{-8}	2.3×10^{-6}
95 GHz	α_s	3.7×10^{-7}	9.1×10^{-5}
	α_b	5.6×10^{-7}	1.1×10^{-4}

2.5 SURFACE CLUTTER SCATTERERS

2.5.1 General Rough Surface Theory

The nature of the reflection of radar signals from a surface depends on the "roughness" of that surface, where the term rough has a very specific definition. It can be seen from Figure 2.6 that the radar signal reflected from two points separated by a difference in height Δh will differ in phase $\Delta\Phi$ by an amount equal to

$$\Delta\Phi = 4\pi\Delta h \sin \Theta/\lambda \qquad (2.28)$$

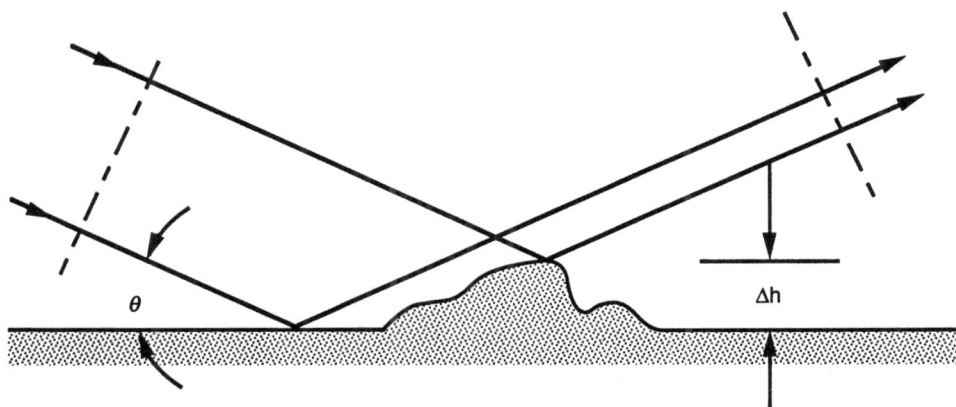

Figure 2.6 Reflected paths for different surface heights. (Adapted from Long, ©1983 by Artech House, Inc.)

where λ is the radar wavelength, and Θ is the incidence angle to the "flat" surface that represents the average of the surface irregularities.

Lord Rayleigh used Equation (2.29) to quantitatively define the terms *rough* and *smooth*. When the two-way path difference between the two scatterers, separated by a root-mean-squared distance Δh, results in a phase difference exceeding one-fourth of a radar wavelength, the surface is defined to be rough; that is,

$$2(\Delta h)_{rms} \sin \Theta \geq \lambda/4 \tag{2.29}$$

The radar signal that is reflected off a rough surface can be considered to have two components, usually called the *specular* (coherent) component and the *diffuse* (noncoherent) component. The specular field phase varies spatially in a deterministic manner, while the diffuse field phase is random and uniformly distributed over the interval 0 to 2π. The mean power density of the sum of noncoherent signals is the *algebraic* sum of the separate power densities, while the total power density of the sum of coherent signals is derived from the *vector* sum of the separate signals and the total power of the summed field [21].

If the illuminated target area is composed of n scattering elements, each of RCS σ_i, the total RCS of the area is given by the expression

$$\sigma = \left| \sum_{i=1}^{n} \sqrt{\sigma_i} e^{j\psi_i} \right|^2 \tag{2.30}$$

where the phase factor $e^{j\psi_i}$ includes phase changes due to reflection and distance.

Long [21] points out that the temporal or spatial average of $\bar{\sigma}$ values is simplified if the separate signals are either totally specular or totally diffuse, where

$$\bar{\sigma} = \sum_{i=1}^{n} \sigma_i \tag{2.31}$$

For totally diffuse reflection, the average $\bar{\sigma}$ will be

$$\bar{\sigma} = \sum_{i=1}^{n} \sigma_i = n\sigma_1 \tag{2.32}$$

if all σ_i values are the same and equal to σ_1. However, when the reflection is specular, the average $\bar{\sigma}$ of σ_i values will given by

$$\bar{\sigma} = \left| \sum_{i=1}^{n} \sqrt{\sigma_i} \right|^2 = |n\sqrt{\sigma_i}|^2 = n^2\sigma_1 \tag{2.33}$$

For example, the radar reflection from the tops of a number of trees in a forest would be expected to be predominantly diffuse. In contrast, the radar return from a wet roadway would be expected to be predominantly specular. Reflections from the ocean surface would contain both specular and diffuse components, depending on the average wave height, wind speed, and wind/wave look direction.

2.5.2 Foliage Backscatter

Values for foliage backscatter coefficients depend on the clutter type, the depression angle, radar frequency, and polarization. Some data indicate only a generally weak frequency dependency for foliage backscatter coefficients. Dependency on depression angle is strong at low angles, below approximately 10°, but is generally flat in the so-called plateau region above 10° until near normal to the average surface plane is reached. Linear cross-polarized backscatter is approximately 5 to 7 dB lower than parallel-polarized backscatter. Cross-polarized circular backscatter is approximately the same magnitude as parallel-polarized circular backscatter [22].

Experimental MMW data indicate that the reflectivity for foliage changes very little when it is cut (i.e., independent of vegetation depth). Seasonal variation in reflectivity also indicates that both the water content and the number of vegetation blades per unit illuminated area strongly influence reflectivity [23].

Radar backscatter from ground vegetation, in particular trees, has been observed at microwave [24] and millimeter [25] wavelengths to consist of a slowly varying component and a higher frequency component. This is particularly obvious in radar returns from wind-blown trees. The frequency components in the return signal depend on wind speed and radar frequency. Similar effects have been observed in sea surface backscatter [26], for which the frequency distributions are a function of sea state, surface winds, and radar frequency.

Backscatter from wind-blown trees has a slowly varying component that produces low frequencies in the frequency, or Doppler, domain. The slowly varying component is associated with the motion of wind-blow tree branches or trunks that are physically large compared to microwave and millimeter-wave radar wavelengths. These scatterers (branches and trunks) produce the major portion of the average RCS at microwave frequencies, but they produce a smaller fraction of the total return at millimeter-wave frequencies. For radars operating at frequencies below L-band, these large branches are the dominant contributors to the RCS. When the radar pulse length and antenna beamwidth are large enough that the cell of resolution contains many trees, the predominant scatterers are the large branches and tree trunks. At the lower radar frequencies, the leaves and twigs are responsible for only a minor portion of the average backscatter from trees, and the leaves and twigs move only a small fraction of a wavelength because of wind motion. The combined result is that the frequency, or Doppler, components are well represented by the Gaussian power spectral density function $P(f)$:

$$P(f) = A \exp(-f^2/2s^2) \qquad (2.34)$$

where

> f = Doppler frequency
> A = Value of $P(f)$ at $f = 0$
> s = Standard deviation of the frequency spectrum

High-frequency components caused by leaf and twig motion of wind-blown trees become evident at S-band and higher frequencies. For radars operating in the K_a-band and M-band, small branches, twigs, and leaves will move through many wavelengths, and, thus, the frequency spread is broader than that seen at X-band. Data collected by Georgia Tech during a four-frequency measurement program [27] show that the spectral content of the backscatter and the functional relationships between scatterers and spectral content are different for frequencies above and below 35 GHz.

The power spectral density function of the high-frequency components resulting from the fast moving portions of the trees follows a Lorentzian form:

$$W(f) = \frac{B}{1 + |f/f_c|^x} \qquad (2.35)$$

where

> B = magnitude near zero frequency
> f_c = corner (half power) frequency, in Hz
> x = 3 for S-, C-, X- and K_u-bands
> x = 2 for K_a- and M-bands

The combined Gaussian and Lorentzian functions are shown in Figure 2.7. Processing of 95-GHz tree backscatter data with a narrow frequency resolution (approximately 1.5 Hz) in the Doppler domain reveals that the slowly moving Gaussian portion of the power spectral density function (PSDF) is approximately 15 dB larger in magnitude than the fast moving Lorentzian portion of the spectrum [28]. Other investigations [23], using less frequency resolution in the data processing, have also suggested that the difference (15 dB) is typical in the microwave region as well as in the millimeter-wave region of the radar frequency domain.

The total area under this combined PSDF must equal the average value of the RCS per unit area, $\bar{\sigma}^0$, which can be determined by processing the measured field

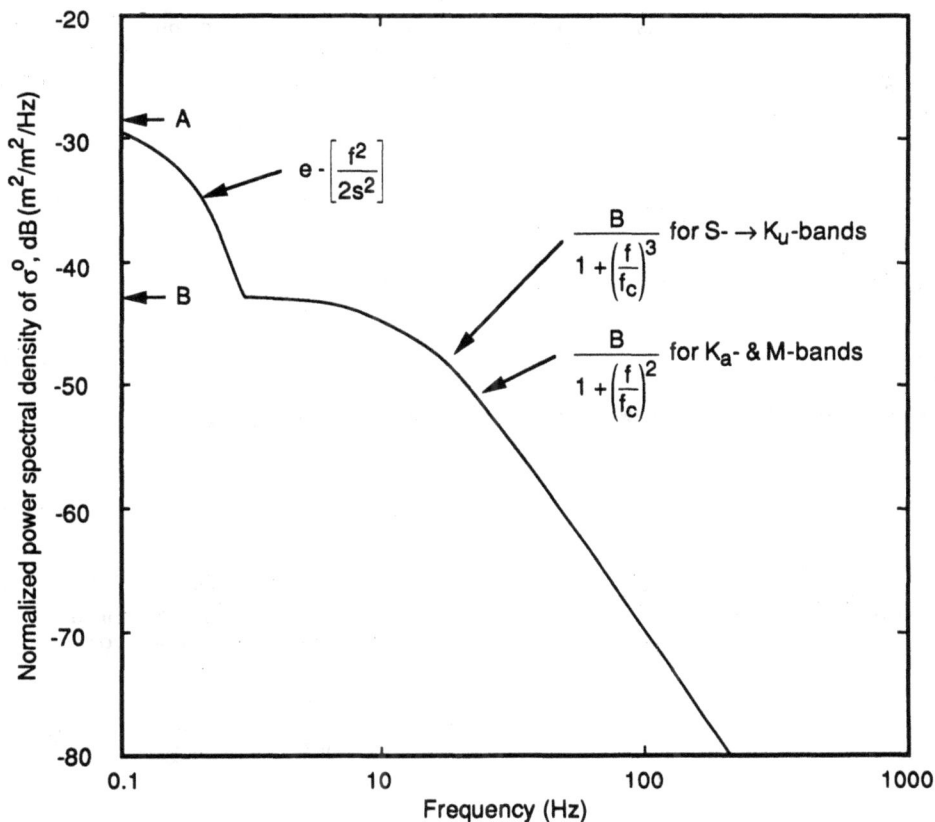

Figure 2.7 Combined Gaussian and Lorentzian function for tree clutter spectrum. (Adapted from Currie, Dyer and Hayes [28].)

data. The area under a Gaussian distribution is given by

$$\int_{-\infty}^{\infty} A \exp(-f^2/2s^2)df = As(2\pi)^{1/2} \tag{2.36}$$

For radar frequencies below S-band, A can be determined very simply from the expression

$$A = \frac{\bar{\sigma}^0}{s(2\pi)^{1/2}} \tag{2.37}$$

For radar frequencies at S-band and higher, the Lorentzian must be included as described. The fractional area δ_1 contained in the spectral far wings (both sides) having amplitude values 15 dB down from the peak value is found to be 0.0085 through integration, as follows

$$\delta_1 = 2 \int_{2.632s}^{\infty} \exp(-f^2/2s^2)df = 0.0085 \qquad (2.38)$$

The total area δ_G under the truncated Gaussian curves is therefore

$$\delta_G = As(0.9915)(2\pi)^{1/2} \qquad (2.39)$$

The Lorentzian portion of the PSDF can be found by integrating

$$\int_{-\infty}^{\infty} \frac{B}{1 + |f/f_c|^x} df \qquad (2.40)$$

and then subtracting out that portion included in the truncated Gaussian section of the distribution. With a change of variables $|f/f_c|^3 = |y|^3$ for S-band through K_u-band and $|f/f_c|^2 = |y|^2$ for K_a-band and M-band, it can be shown that [29]

$$f_c \int_{-\infty}^{\infty} \frac{B}{1 + |y|^3} dy = f_c B \frac{4\pi}{3(3)^{1/2}} \qquad (2.41)$$

and

$$f_c \int_{-\infty}^{\infty} \frac{B}{1 + |y|^2} dy = f_c B \pi \qquad (2.42)$$

The area under the Lorentzian portion of the distribution common with the Gaussian portion can be removed by subtracting the area (for K_u- and lower bands) defined by

$$\delta_2 = 2f_c \int_0^Y \frac{B}{1 + |y|^3} dy \qquad (2.43)$$

$$\delta_2 = 2f_c B \left[\frac{1}{6} \ln \left(\frac{(1 + Y)^2}{1 - Y + Y^2} \right) + 3^{-1/2} \tan^{-1} \left(\frac{2y - 1}{\sqrt{3}} \right) \right] \qquad (2.44)$$

where $y = f_1/f_c$ and $f_1 = $ the value of frequency when the Gaussian portion has been reduced to 15 dB below the peak value, and the Gaussian portion intersects with the Lorentzian portion. Values for f_1, f_c, and Y are given in Table 2.6.

Using the values of Y, f_1, and f_c from Table 2.6, it can be shown that δ_2 will amount to about 18.5% of the total area under the Lorentzian curve. The area under the truncated Lorentzian curve δ_L is thus given by the expression

$$\delta_L = \int_{-\infty}^{\infty} \frac{B}{1 + |f/f_c|^3} \, df = -\delta_2 \qquad (2.45)$$

Equation (2.45) will be used for radar frequency bands S, C, X, and K_u.

In a similar manner, the area of the Lorentzian to be removed for the K_a-band and M-band data is found from the expression

$$\delta_2 = 2f_c \int_0^Y \frac{B}{1 + y^2} \, dy = 2f_c B[\tan^{-1} Y]^{f_1/f_c} \qquad (2.46)$$

The average cross section per unit area can now be equated with the area under the PSDF, namely

$$\bar{\sigma}^0 = A[(2\pi)^{1/2}s - \delta_1] + B\left[f_c \frac{4\pi}{3(3)^{1/2}} - \delta_2\right] \qquad (2.47)$$

Since $A/B = 15$ dB, or 31.62, the values of δ_1 and δ_2 can be used to rewrite Equation (2.47) as

$$\bar{\sigma}^0 = A[0.9915(2\pi)^{1/2}s] + B[1.90707 f_c] \qquad (2.48)$$

Table 2.6
Parameter Values for Equation (2.44)

Radar Band	f_1 (Hz)	f_c (Hz)	Y
S	0.26	1.2	0.21883
C	0.95	4.2	0.22619
X	3.00	13.3	0.22556
K_u	5.00	22.0	0.22727
K_a	9.00	53.0	0.16981
M	15.00	140.0	0.10714

Then, for radar frequency bands S, C, X, and Ku,

$$\bar{\sigma}^0 = A[2.4853s + 0.062326f_c] \qquad (2.49)$$

for K_a-band,

$$\bar{\sigma}^0 = A[2.4853s + 0.088716f_c] \qquad (2.50)$$

and for M-band,

$$\bar{\sigma}^0 = A[2.4853s + 0.092603f_c] \qquad (2.51)$$

where

$\bar{\sigma}^0$ is in dB (m^2/m^2)
A is in dB $(m^2/m^2/Hz)$

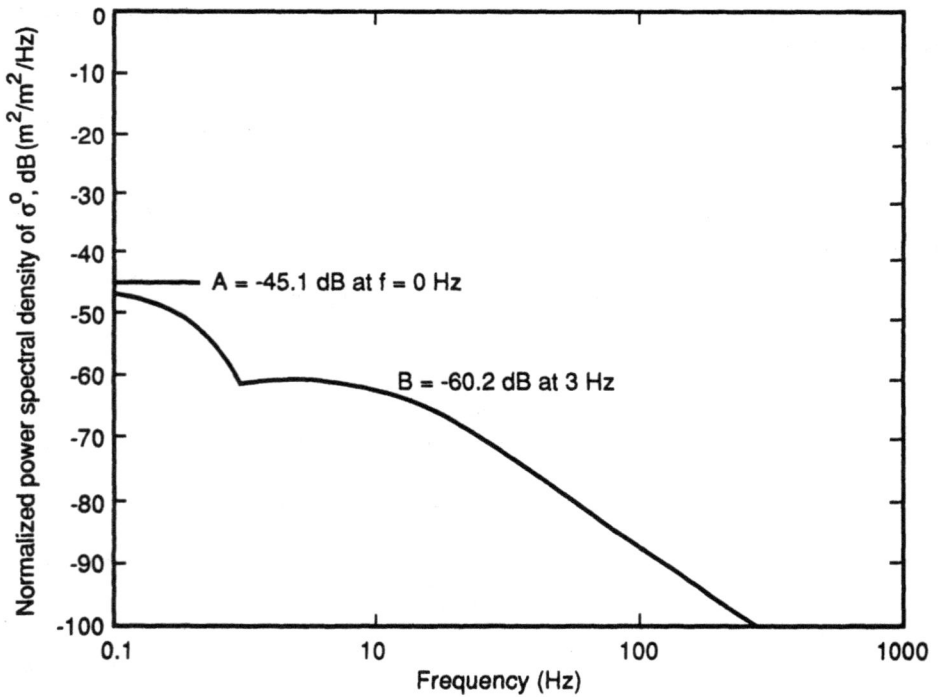

Figure 2.8 Power spectral density function for wind-blown tree backscatter at X-band, 15 mph wind speed. (Adapted from Hayes et al. [24].)

s is in Hz
f_c is in Hz

Values for A (and B) can be determined by using values of σ^0 derived from the equations in Table 3.14, which were developed from experimental data [30].

A power spectral density function can be derived by using empirical values for s and Equations (2.49), (2.50), and (2.51). An example of such a PSDF for X-band is shown in Figure 2.8 [24].

2.5.3 Backscatter From Snow-Covered Ground

The radar backscatter from surfaces such as snow-covered ground differs from other surfaces in that it often involves the reflection of electromagnetic energy from two or more layers of materials, each of which may have dielectric properties that are distinctly different from the others. In order to analyze this situation, consider Figure 2.9, which depicts a cross section of a typical snowpack geometry with a layer of snow d meters thick, illuminated by a radar signal along a line of sight oriented at an angle Θ to the surface normal. The specular reflection from the snow surface will also be oriented at an angle Θ from the surface normal, while the refracted energy propagating through the snow down to the ground surface will be oriented at an angle Θ' with respect to the surface normal.

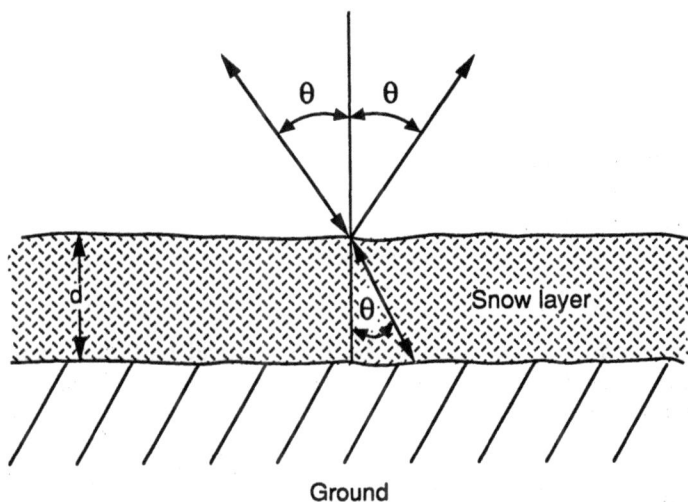

Figure 2.9 Cross section of a typical snowpack geometry. (Adapted from Stiles and Ulaby [31].)

The specific physical properties of the snowpack, along with their associated symbols, that affect both the reflection from the top of the snowpack and the propagation through the snowpack include

Snow surface roughness
Depth of the snow, d
Snow density, r
Snow water equivalent, W
Snow wetness, m_v
Snow crystalline structure
Snow stratification
Snow inhomogeneities
Ground surface roughness
Soil moisture, m_s
Snow physical temperature

where the water equivalent is the equivalent depth of water resulting from the melting of the entire snow layer (i.e., the product of snow depth and snow density) [31].

The dielectric properties of snow are made all the more complicated by their structural dependencies, such as water content, crystal structure, and packing profile. As a result, the bulk dielectric properties of the snowpack tend to be modeled using mixing formulas consisting of two or more components. The Weiner formula for the dielectric constant for dry snow k_{ds} is representative of the genre:

$$\frac{k_{ds} - 1}{k_{ds} + F} = \rho_i \frac{k_i - 1}{k_i + F} + (1 - \rho) \frac{k_0 - 1}{k_0 + F} \qquad (2.52)$$

where

ρ_i = volumetric fraction of ice
k_0 = dielectric constant of air
k_i = dielectric constant of ice
k_{ds} = dielectric constant of dry snow
F = form number

The form number F may be equal to zero for vertical particles, equal to two for spherical particles, and equal to infinity for elongated, horizontal particles [32]. The loss tangent $\tan \delta_s$ for dry snow was modeled by Edgerton et al. as a polynomial:

$$\tan \delta_s = 0.756 \times 10^{-9} \exp\left[0.0574\, T + \frac{3.277}{A^2} - \frac{4.67}{A^3} + \frac{1.55}{A^4} \right] \qquad (2.53)$$

where $A = (273.5 - T)$ and T is the kelvin temperature ($T < 273$ K) [33].

The attenuation coefficient α (in dB/m) of an electromagnetic wave through a snowpack due to absorption is given by the expression

$$\alpha = \frac{17.36\pi}{\lambda} \left\{ \frac{k_r'}{2} \left[\left(1 + \left(\frac{k_r''}{k_r'} \right)^2 \right)^{1/2} - 1 \right] \right\}^{1/2} \qquad (2.54)$$

where

λ = radar wavelength

k_r' = real part of the relative dielectric constant of the snow

k_r'' = imaginary part of the relative dielectric constant of the snow

Computation of the reflectivity σ^0 from snow may be performed using the so-called multiple-stream method [34]. The method is based on treating the scattering intensity from a collection of identical particles with a known scattering phase function, which is taken from Ulaby et al. [35] and outlined here. The snow is modeled first as a thin layer of scatterers of optical depth $\Delta\tau$. For a sufficiently thin layer, single scattering is assumed. Forward and backward scattering functions may be evaluated in terms of the given phase function. Multiple scattering effects resulting from two thin layers are then established by tracing each "bounce." In matrix form, the backward scattering matrix $\mathbf{S}$ and the forward scattering matrix $\mathbf{T}$ for the combined layers can be written in terms of the corresponding backward and forward matrices $\mathbf{S}_{1,2}$ and $\mathbf{T}_{1,2}$ of layers 1 and 2 as the expressions

$$\mathbf{S} = \mathbf{S}_1 + \mathbf{T}_1\mathbf{S}_2(\mathbf{I} - \mathbf{S}_1\mathbf{S}_2)^{-1}\mathbf{T}_1 \qquad (2.55)$$

and

$$\mathbf{T} = \mathbf{T}_2(\mathbf{I} - \mathbf{S}_1\mathbf{S}_2)^{-1}\mathbf{T}_1 \qquad (2.56)$$

where $\mathbf{I}$ is the unit matrix. The $\mathbf{S}$ and $\mathbf{T}$ matrices now define the backward and forward matrices for a layer of snow $2\Delta\tau$ thick. This process is repeated for a layer $4\Delta\tau$ thick, and so forth. An appropriate air-snow boundary and a ground plane must also be added.

2.6 ATTENUATION

Signal attenuation, previously defined in Section 1.3, degrades a radar's signal-to-noise ratio, thus reducing the effectiveness of all aspects of radar performance. At millimeter-wave frequencies, atmospheric attenuation coefficients tend to be significantly larger than those at microwave frequencies for all varieties of naturally occurring atmospheric constituents and hydrometeors. Attenuation due to manmade obscurant materials, which have been tailored for use against electro-optic sensors, are ineffective against radar sensors at either microwave or millimeter-wave frequencies. Even attenuation caused by soil ejecta created by explosions or caused by road dust is short lived, lasting only as long as appreciably large particles of dirt remain in the air along the radar line of sight.

This section will address the attenuation of electromagnetic waves, especially at millimeter-wave frequencies, created by clear air (normally occurring gaseous molecules), hydrometeors (fog, rain, and snow), and manmade obscurants (smoke, aerosols, and road dust).

2.6.1 Clear Air Attenuation

The standard dry atmosphere consists of approximately 78.088% oxygen, 20.949% nitrogen, 0.93% argon, and 0.03% other molecules, including water vapor. Atmospheric attenuation in the millimeter-wave region is dominated, however, by the air's water vapor and oxygen content. Ozone also has significant signal absorption effects in the millimeter-wave region, but ozone exists primarily at high elevations, so the effect is of little importance to tactical millimeter-wave seekers. The electrically polar water molecule has a strong dipole moment, and the nature of the energy absorption is due to transitions between molecular rotational states. These states are closely spaced together in three specific regions within the millimeter-wave region: 22, 183, and 323 GHz [36].

Oxygen absorption of electromagnetic energy is caused by a variety of physical molecular transition states. The oxygen molecule is diatomic and has no dipole moment. There are several resonance lines because of rotational state transitions, but these mechanisms are weak and very much dominated by the absorption due to water vapor at nearby frequencies. Some energy absorption in the millimeter-wave region is due to oxygen and is caused by magnetic dipole transitions. The most significant oxygen absorption resonances are around 60 GHz (many lines) and at 118 GHz (single line) because of transitions between triplet components of the oxygen rotational ground energy state [37].

The attenuation coefficient values (all in dB/km), α_1 due to the oxygen in the atmosphere, α_2 due to the water vapor (at 22 GHz), and α_3 due to the water vapor (above 22 GHz) in the millimeter-wave portion of the electromagnetic spectrum, modified from Van Vleck [38], can be mathematically described by the expressions

$$\alpha_1 = \frac{0.34}{\lambda} \left[\frac{\Delta \nu_1}{1/\lambda^2 + \Delta \nu_1^2} + \frac{\Delta \nu_2}{(2 + 1/\lambda)^2 + \Delta \nu_2^2} + \frac{\Delta \nu_2}{(2 - 1/\lambda^2)^2 + \Delta \nu_2^2} \right] \quad (2.57)$$

$$\alpha_2 = \frac{3.5\rho \times 10^{-3}}{\lambda^2} \left[\frac{\Delta\nu_3}{(1/\lambda - 1/1.35)^2 + \Delta\nu_3^2} + \frac{\Delta\nu_3}{(1/\lambda - 1/1.35)^2 - \Delta\nu_3^2} \right] \quad (2.58)$$

$$\alpha_3 = \frac{0.05\rho\Delta\nu_4}{\lambda^2} \quad (2.59)$$

where λ is the radar wavelength, ρ is the absolute humidity in g/m^3, and the $\Delta\nu_i$ are linewidth factors [39]. Figure 2.10 depicts the attenuation coefficient α as a function of radar frequency from the microwave region to the near millimeter-wave region, at sea level and at 4 km altitude [37]. It is readily apparent from this figure that radars designed to operate in the millimeter region of the electromagnetic spectrum will have operating frequencies in the relative minima of the attenuation/frequency curve. These are seen to occur at 35, 94, 140, and 220 GHz. When maximum range is important and covert operation is not, these so-called "radar windows" represent the natural selection for operating frequency, subject to other system constraints.

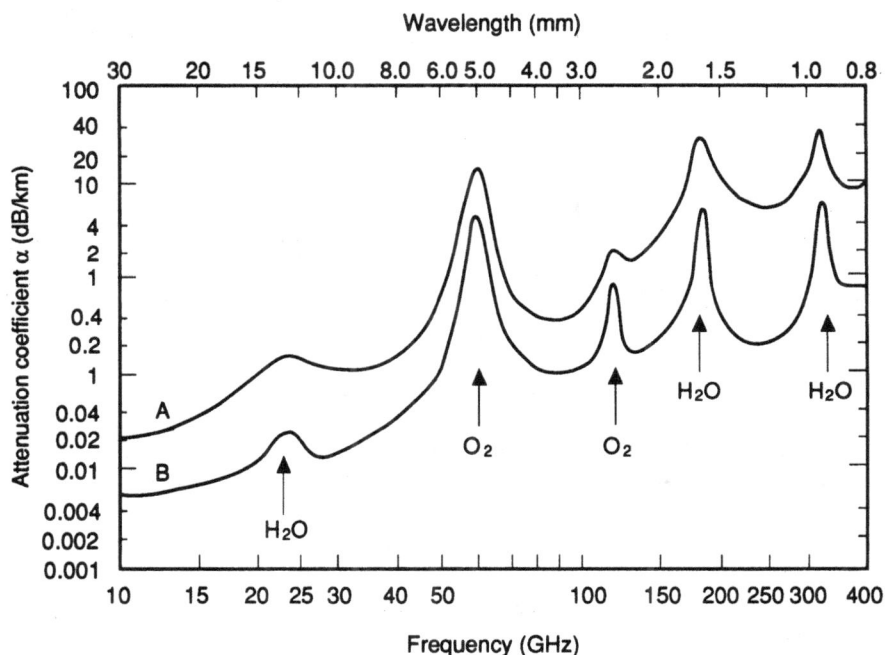

Figure 2.10 Average atmospheric absorption of millimeter waves. Curve A represents sea-level attenuation coefficient values, and curve B represents attenuation coefficient values at 4 km altitude. (Adapted from Wiltse, ©1981 by Academic Press [37].)

2.6.2 Hydrometeor Attenuation

The class of meteorological attenuation of interest to radar engineers includes fog, clouds, and haze; rain and snow; and hail. Fog, rain, and snow attenuation will be addressed. Hail occurs infrequently and will not be covered here.

2.6.2.1 Fog and Haze Attenuation

One of the perceived operational advantages for using millimeter-wave radars rather than infrared or visible wavelength sensors is the ability to penetrate more effectively through water suspended in the atmosphere (i.e., fog and haze). When all types of hydrometeors are included, this capability of millimeter-wave radar systems is often referred to as an "all weather" capability.

Clearly, even millimeter-wave systems will be subject to signal attenuation when passing through an atmosphere containing water in the form of very small droplets (drop diameters less than 10 μm for radiation fogs and up to 100 μm for advection fogs). The small size of fog droplets allows the use of Rayleigh approximations for calculating attenuation levels. The normalized one-way attenuation coefficient α due to fog can be described mathematically by the expression [8]

$$\alpha = \frac{81.86 \, M \, \text{Im}(-K)}{\lambda \rho} \tag{2.60}$$

where α is in units of dB/km, λ is the radar wavelength in mm, M is the liquid water content per unit volume of fog in g/m^3, ρ is the density of water in g/cm^3 (essentially 1.00 g/cm^3 from 0°C to 40°C), and $\text{Im}(-K)$ is the absorption coefficient, where

$$K = (m^2 - 1)/(m^2 + 2) \tag{2.61}$$

The complex valued index of refraction m can be defined in terms of the complex valued dielectric constant for liquid water ε_c using the expression

$$m^2 = \varepsilon_c = \varepsilon_1 - j\varepsilon_2 \tag{2.62}$$

where ε_1 and ε_2 are the real and imaginary components, respectively, of the dielectric constant [40].

An empirical relationship between the optical visibility V in fog and the fog liquid water density M has been given by Eldridge as

$$V = 0.024 \, M^{-0.65} \tag{2.63}$$

where V has units of km and M has units of g/m^3 [7]. Equation (2.63) is thought to be valid for radiation fogs containing drops whose diameters are less than 10 μm. For the case of advection fogs having droplets whose diameters exceed 100 μm, Eldridge recommends a coefficient of 0.017 in Equation (2.63).

Since the complex index of refraction for water is very temperature-dependent between 0°C to 40°C, it follows that the attenuation will also be temperature-dependent. Altschuler's analysis of measured millimeter-wave fog attenuation data indicates a straightforward relationship relating attenuation coefficient α to the radar wavelength and temperature T [41]:

$$\alpha/M = -1.347 + 0.372\lambda + \frac{18.0}{\lambda} - 0.022T \qquad (2.64)$$

where

α/M = attenuation, in dB/km/g/m^3
λ = radar wavelength, in mm

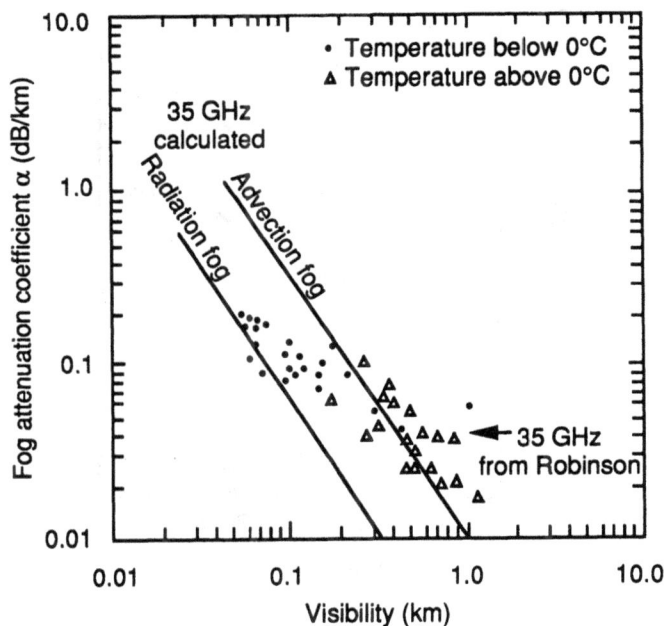

Figure 2.11 Measured and calculated fog attenuation coefficients at 35 GHz. (Adapted from Robinson, ©1955 by IEEE [42].)

T = temperature, in °C
M = fog density, in g/m^3

The use of Equation (2.31) should be restricted to radar wavelengths between 3 mm and 3 cm and a temperature range between −8°C and +25°C.

Figure 2.11 depicts calculated and measured values of both radiation and advection fog attenuation (dB/km) as a function of visibility (in km) [42].

Figure 2.12 shows calculated attenuation data versus temperature and radar wavelength. The water vapor contribution was linearly extrapolated from absolute humidity without temperature corrections, and fog contributions were calculated in the Rayleigh limit for water vapor at a temperature of 24°C. As a result, fog components may be up to 50% in error [43].

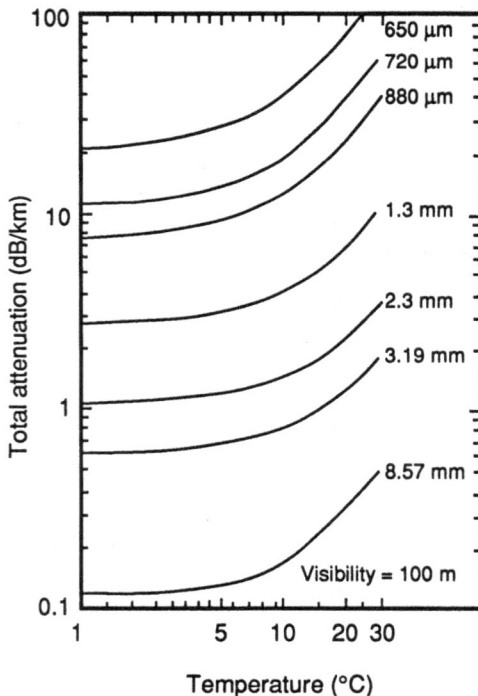

Figure 2.12 Fog attenuation as a function of temperature and wavelength. (Adapted from Kulpa and Brown [43].)

2.6.2.2 Rain Attenuation

The attenuation of a radar signal due to rain has been extensively researched, both experimentally and analytically. Rain attenuation has been determined to depend heavily on drop size distribution, which in turn can depend on rainfall rate, geographic location, and time of the year. Analytical approaches are based on the fact that a water drop is essentially a conducting sphere when the drop diameter is less than approximately 3 mm. The drop size distributions most often used to model rain attenuation are the Marshall-Palmer distribution and the Laws-Parsons distribution.

Based on actual recorded data, rain attenuation coefficient data have been empirically modeled by Georgia Tech researchers and others in both the microwave and the millimeter-wave regions as a two-parameter monomial:

$$\alpha = A R^B \qquad (2.65)$$

where α is in dB/km, R is the rainfall rate in mm/hr, and A and B are taken from Table 2.7. Figure 2.13 shows curves of the parameters A and B as a function of frequency based on measured attenuation data [44]. Values for the parameters A and B can then be selected from these curves for a specified radar frequency. Note the strong frequency dependency on the parameter A, but the relatively weak frequency dependency on the parameter B.

The effect of temperature on attenuation is involved through the refractive index of the rain drop. Its effect at millimeter-wave frequencies will not exceed $\pm 20\%$

Table 2.7
Predicted Rainfall Attenuation Coefficients as a Function of Rainfall Rate
(Adapted from Currie, Dyer, and Hayes [12])

	α (dB/km) = AR^B			
	10 GHz	35 GHz	70 GHz	95 GHz
R (mm/hr)	A = 0.009 B = 1.16	A = 0.273 B = 0.985	A = 0.634 B = 0.868	A = 1.6 B = 0.64
5	0.05	1.33	2.56	4.48
10	0.13	2.64	4.68	5.98
15	0.21	3.93	6.65	9.05
20	0.30	5.22	8.54	10.88
25	0.39	6.50	10.36	12.55

72

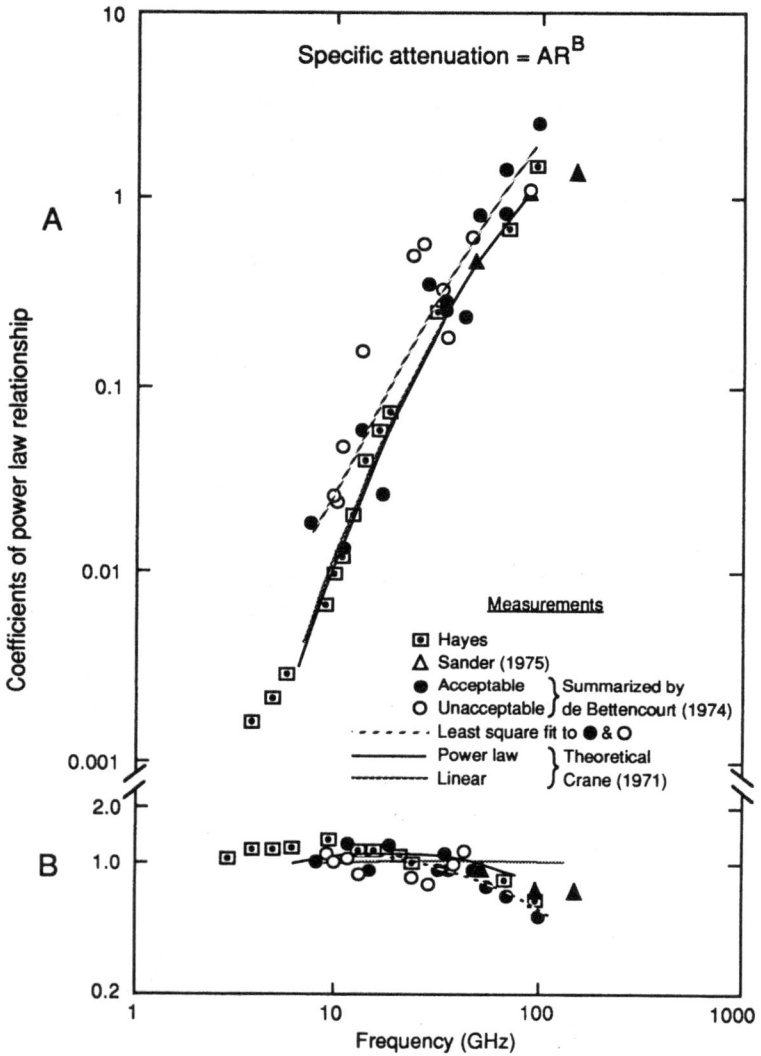

Figure 2.13 Coefficients of power law relationships between attenuation and rainfall rate.

between 10°C and 30°C, and is usually much less [46]. The values of the index of refraction for water as functions of radar wavelength and temperature are listed in Table 2.8 [40]. Computation of attenuation coefficient values may be performed using the appropriate data from Table 2.8 and the techniques outlined in [46].

Both signal backscatter and signal attenuation phenomena demonstrate polarization dependencies when the rain drops become nonspherical (i.e., when their diameters exceed approximately 3 mm). This condition becomes more prevalent during heavy rainfall, which tends to include the larger drop sizes. Furthermore, with a blowing wind, the drops' axes of symmetry become canted with some mean orientation angle with respect to the vertical. Depolarization of electromagnetic signals by hydrometeors is considered to be due to their nonsphericity. Horizontally polarized signals are attenuated more strongly than are vertically polarized signals, since the rain drops tend to "flatten" as they fall, especially the larger ones. For linear signal polarization, the depolarization effect is due to the difference between the signal polarization angle and the mean canting angle of the hydrometeors [47].

A heuristic description of a nonspherical rain drop falling at a nonzero canting angle leads to the conclusion that a differential amplitude attenuation and a differential phase shift will exist with respect to, for example, horizontal and vertical

Table 2.8

Index of Refraction for Water as Functions of Wavelength and Temperature
(Adapted from Gunn and East [40])

	Temp (°C)	Wavelength (cm)		
		3.21	1.24	0.62
n	20	8.14	6.15	4.44
	10	7.80	5.45	3.94
	0	7.14	4.75	3.45
	−8	6.48	4.15	3.10
k	20	2.00	2.86	2.59
	10	2.44	2.90	2.37
	0	2.89	2.77	2.04
	−8		2.55	1.77
$|K|^2$	20	0.9275	0.9193	0.8926
	10	0.9282	0.9152	0.8726
	0	0.9300	0.9055	0.8312
	−8		0.8902	0.7921
$Im(-K)$	20	0.0188	0.0471	0.0915
	10	0.0247	0.0615	0.1142
	0	0.0335	0.0807	0.1441
	−8		0.1036	0.1713

polarizations [48,49]. Likewise, a polarization-dependent backscatter and propagation can be expected because the rain drops have a preferred orientation and are nonspherical, conducting scatterers.

Signal propagation depolarization can be quantified by differential attenuations and differential phase shifts. Thus, a propagating signal may be attenuated more with respect to one polarization state than the orthogonal polarization state. Likewise, the signal may be retarded in phase by different amounts. The net effect is a change in the polarization state of the transmitted pulse, so that the electromagnetic characteristics of the radar pulse may become distinctly different at the range of the target from those that left the antenna. The effect can be defined as a change in the polarization ellipticity of the signal or simply a phase rotation under some conditions. Since most radars operate with a common antenna for transmission and reception (i.e., monostatic operation), the depolarization effected during transit to the target will be further degraded as the reflection from the target cell propagates back toward the radar through an identical medium.

The differential attenuation coefficient $\Delta\alpha = \alpha_h - \alpha_v$ and the normalized differential attenuation coefficient $\Delta\alpha_n = (\alpha_h - \alpha_v)/\alpha_v$ are the pertinent depolarization parameters for horizontal (h) and vertical (v) polarizations. For the assumption of constant rain rate and uniform rain drop axis orientation, the differential attenuation has a maximum at 30 GHz for high rain rates, while the normalized differential attenuation has a maximum at 5 GHz and a secondary maximum at 20 GHz. The differential phase shift $\Delta\Phi = \Phi_h - \Phi_v$ becomes very large at 20 GHz for high rain rates, while at frequencies higher than 35 GHz the differential phase shift becomes negative! Note that around 35 GHz the differential phase shift is bounded in a very small range for all possible rain rates.

Figure 2.14 shows the calculated differential attenuation (horizontal polarization − vertical polarization) as a function of rainfall rate and radar frequency. Figure 2.15 shows the calculated differential phase shift (horizontal polarization − vertical polarization) as a function of rainfall rate and radar frequency [50]. Table 2.9 lists calculations for the parameters A and B from Equation (2.65) for both horizontally and vertically polarized signals and several radar microwave and millimeter-wave frequencies. Note that these values are fairly close to those listed in Table 2.2 [51].

One method of computing more realistic values for depolarization parameters is to assume that the rain drop canting angles are nonzero, but the same. The following cross-polarization factors are then defined for transmitted orthogonal (horizontal and vertical) electric field polarizations:

$$\text{XPI}_{hh} = 20 \log_{10} \left| \frac{\Delta E_h}{E_h} \right| \qquad \text{XPI}_{vv} = 20 \log_{10} \left| \frac{\Delta E_v}{E_v} \right|$$

$$\text{XPD}_{hv} = 20 \log_{10} \left| \frac{\Delta E_v}{E_h} \right| \qquad \text{XPD}_{vh} = 20 \log_{10} \left| \frac{\Delta E_h}{E_v} \right| \qquad (2.66)$$

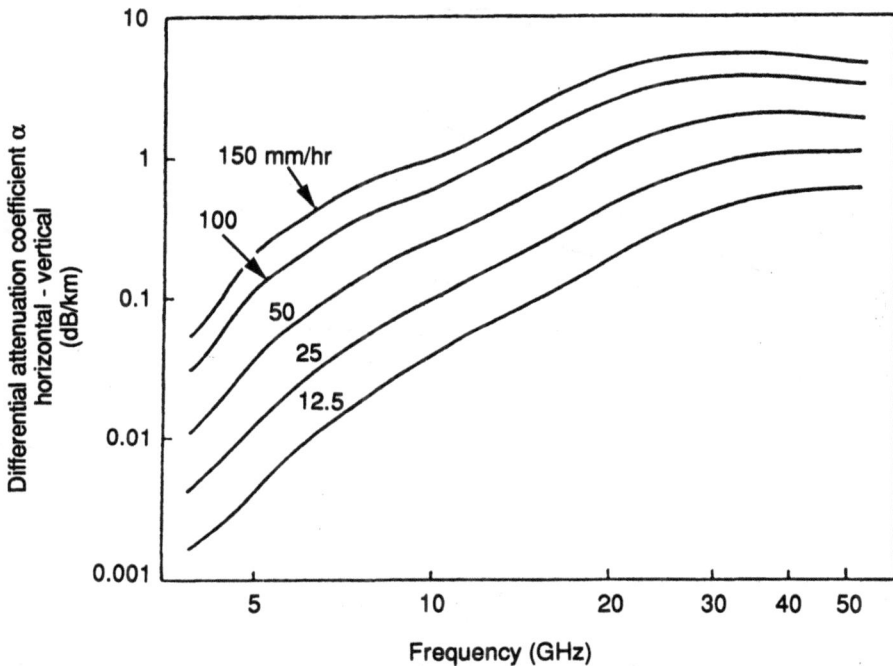

Figure 2.14 Calculated differential attenuation coefficient (horizontal polarization − vertical polarization) as a function of frequency for different rainfall rates. (Adapted from Oguchi and Hosoya [50].)

where

E_h and E_v	are the copolarized received electric fields
ΔE_h	is the cross-polarized electric field in the direction E_h transfered from E_v
ΔE_v	is the cross-polarized electric field in the direction E_v transfered from E_h
XPI_{hh} and XPI_{vv}	are the cross-polarization isolation factors
XPD_{hv} and XPD_{vh}	are the cross-polarization discrimination factors

When a constant canting angle assumption is made, these equations indicate that $\Delta E_h = \Delta E_v$, so that $XPD_{hv} = XPI_{hh}$ and $XPD_{vh} = XPI_{vv}$. When this assumption is relaxed (i.e., the drop canting angles are distributed and the drop shapes are arbitrary), ΔE_h is not usually equal to ΔE_v, and $XPD_{hv} \neq XPI_{hh}$ and $XPD_{vh} \neq XPI_{vv}$ [48].

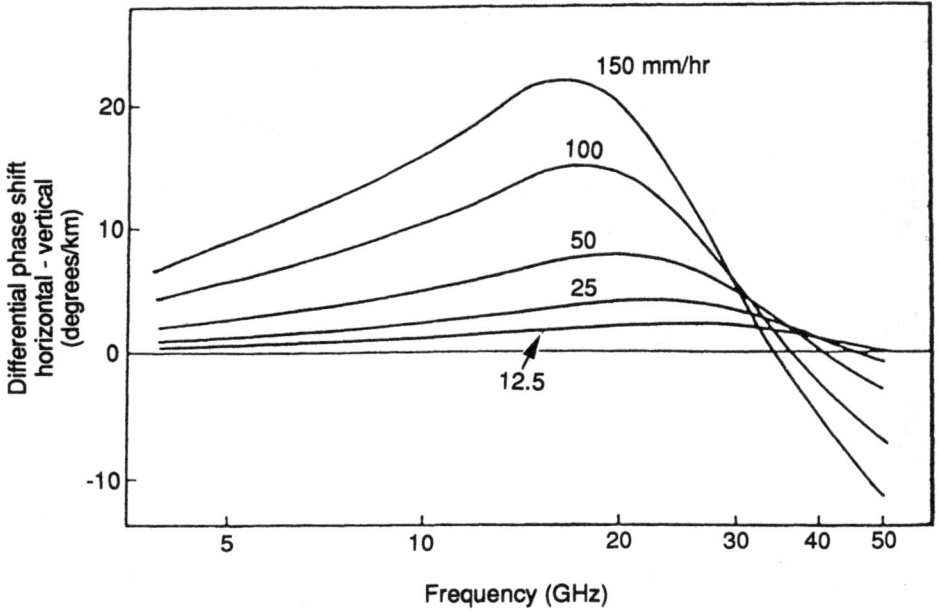

Figure 2.15 Calculated differential phase shift (horizontal polarization − vertical polarization) as a function of frequency for different rainfall rates. (Adapted from Oguchi and Hosoya [50].)

Table 2.9
Parameters A and B for Attenuation Coefficient a for Horizontal and Vertical Polarizations at 18°C
(Adapted from the Commission of the European Communities [51])

Frequency (GHz)	α (dB/km) $= A R^B$			
	A		B	
	Horizontal	Vertical	Horizontal	Vertical
10	0.010	0.009	1.276	1.264
15	0.037	0.035	1.154	1.128
35	0.263	0.233	0.979	0.963
95	1.094	1.033	0.748	0.748

2.6.2.3 Snow Attenuation

The attenuation due to falling snow is far more difficult to model analytically since snow flakes can assume a wide variety of shapes, depending on the manner in which the snow was formed. Snow may take the form of aggregates of ice crystals in the shapes of flakes, crystals, needles, or even pellets. Snow flakes may range in size from a few millimeters in diameter up to as much as several centimeters. Furthermore, they can be described as either "dry" or "wet," depending on the free water adhering to the flakes as they fall. Thus, the attenuation due to falling snow has tended to be described empirically.

The snow mass concentration X can be related to equivalent rainfall rate R_e using the expression

$$R_e = 1000 \, Xv/\rho \qquad (2.67)$$

where X has units of g/m^3, R_e has units of mm/hr, ρ is the mass denisty of water in g/mm^3, and v is the fall velocity of the snow (modified from [52]).

The following expression has been proposed for snowfall attenuation coefficient α in the microwave region:

$$\alpha = 0.00349 \frac{R^{1.6}}{\lambda^4} + 0.00224 \frac{R}{\lambda} \qquad (2.68)$$

where α is in decibels per kilometer, λ is in centimeters, and the snowfall rate R is in millimeters of water per hour [40]. Figure 2.16 shows calculated attenuation coefficient values (one-way at 0°C) as a function of both radar frequency (from S-band through K_a-band) and equivalent rainfall rate. Figure 2.17 depicts both theoretical and measured values for snowfall attenuation coefficients at 35 GHz [53].

2.6.3 Foliage Attenuation

A tree consists of a trunk, branches, twigs, and leaves. Depending on the wavelength of the electromagnetic energy, each portion of a tree will cause varying amounts of attenuation and scattering. Branches and twigs will cause effects somewhere between those of leaves and trunks. Consider a leaf as a polygon surface whose dimensions can be described in 3-space. The material is considered to be a lossy dielectric consisting of water with minerals. The model of a tree, then, will consist of a collection of randomly oriented leaves and the air between the leaves.

Since the leaves are randomly oriented, we must ask what the scattering from such a surface is. Knott calls this the "tumble average" bistatic scattering cross section $\langle\sigma\rangle$ of a leaf, where

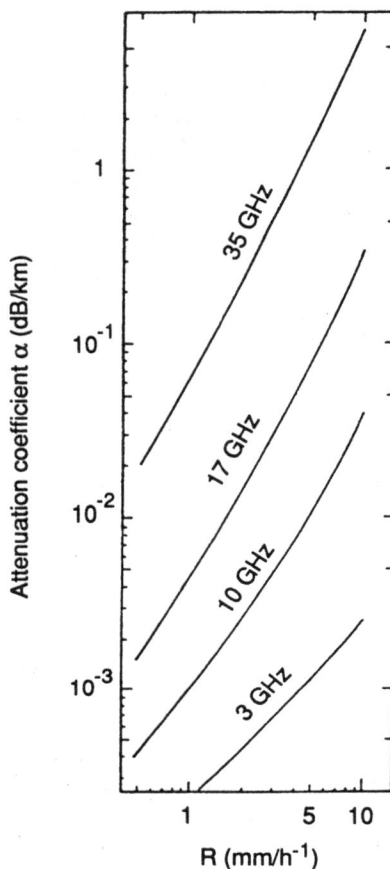

Figure 2.16 Calculated snow attenuation coefficient (one-way) as function of equivalent rainfall rate for different frequencies. (Adapted from Sauvageot [9].)

$$\langle\sigma\rangle = \frac{1}{4\pi}\int\int\sigma\,d\Omega \qquad (2.69)$$

and the leaf itself is treated as a dielectric slab [54].

The Tennessee Valley Authority (TVA) Forestry Service conducted a study in Tennessee, Georgia, and North Carolina to determine the leaf content of trees and how the foliage affected the water table levels, soil erosion, wildlife, and general forest health. A detailed sampling process of leaves per branch, branches per tree, branches for a tree trunk diameter, and types of trees was undertaken, which determined that the number of leaves on a large class of trees can be related to the di-

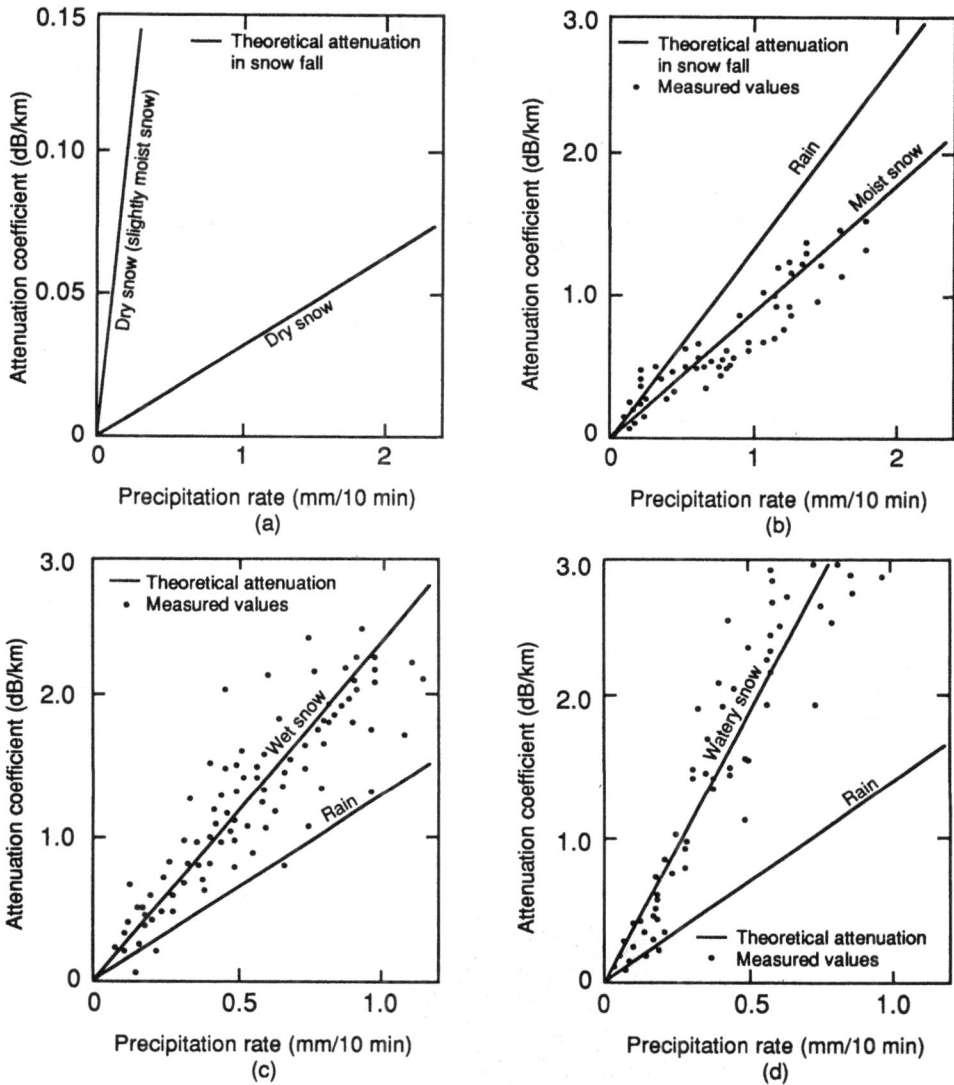

Figure 2.17 Theoretical and measured attenuation coefficients for snow and rain at 35 GHz. (Adapted from Nishitsuji and Matsumato [53].)

ameter of the trunk as measured at breast height (dbh). Actually this is the trunk diameter 4.5 ft above the ground. In general, the number of leaves N is given by

$$\log N = A + B \log (\text{dbh}) \tag{2.70}$$

where A and B are parameters specific to different classes of trees. It was also noted in the TVA study that the number of leaves did not vary with leaf crown class, nor with location in the upper or lower sections of the crown [55].

The geometrical shapes of the foliage tree crown must be consistent with "Mother Nature's" requirement that leaves see maximum sunlight in order to perform the photosynthesis process. This process is so important that leaves have a built-in twisting mechanism that allows the leaf to rotate and capture more sunlight as a function of the time of day. A term often used to quantify geometric shape is the leaf index, defined as

$$\text{Leaf index} = \frac{\text{Surface area of the leaves}}{\text{Drip line area on ground}} \tag{2.71}$$

as shown in Figure 2.18. In Forestry Service terms, this index varies from 4 to 10. A representative shape for the foliage crown is an ellipsoid of revolution.

The effective dielectric constant ε_{eff} for a mixture of two materials is given by the ratio of volumes of materials, i.e.,

$$\varepsilon_{\text{eff}}[V_{\text{air}} + V_{\text{leaf}}] = \varepsilon_{\text{air}}V_{\text{air}} + \varepsilon_{\text{leaf}}V_{\text{leaf}} \tag{2.72}$$

where V is the volume of the appropriate material, ε_{air} is equal to 1, and $\varepsilon_{\text{leaf}}$ is a complex quantity. The real part of ε_{eff} will cause some ray bending or refraction when the angle of the electromagnetic wave is not normal to the foliage surface. The imaginary part of ε_{eff} will cause absorption, so that

$$\varepsilon'_{\text{eff}}V_{\text{tree}} = 1 \times V_{\text{air}} + \varepsilon'_{\text{leaf}}V_{\text{leaf}} \tag{2.73}$$

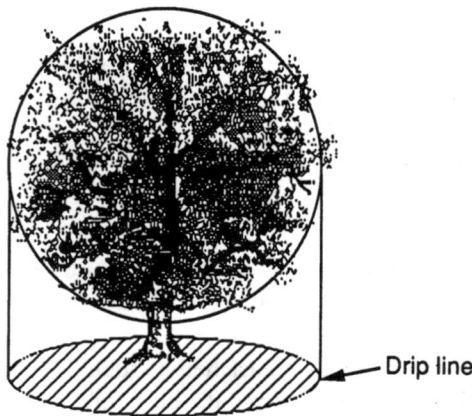

Figure 2.18 Tree drip line.

and

$$\varepsilon''_{\text{eff}} V_{\text{tree}} = \varepsilon''_{\text{leaf}} V_{\text{leaf}} \tag{2.74}$$

where ε' and ε'' are the real and imaginary component of ε, respectively. Solving for $\varepsilon''_{\text{eff}}$ yields the expression

$$\varepsilon''_{\text{eff}} = \varepsilon''_{\text{leaf}} \frac{V_{\text{leaf}}}{V_{\text{tree}}} \tag{2.75}$$

The dielectric constant for a leaf $\varepsilon''_{\text{eff}}$ is chosen to be between the values of ε'' for fresh water and ε'' for salt water (36/1000 salinity). The attenuation coefficient α through a dielectric slab, as shown in most any microwave text book, is given by the expression

$$\alpha = 4.343 \frac{2\pi\varepsilon''_{\text{eff}}}{\lambda} = 4.343 \frac{2\pi\varepsilon''_{\text{leaf}}}{\lambda} \frac{V_{\text{leaf}}}{V_{\text{tree}}} \tag{2.76}$$

The forward scattering and absorption due to leaf foliage can then be determined from the calculations of the bistatic scattering cross section at the forward scattering angles. In addition to the foliage, there is also scattering and absorption by the branches and trunks of trees. Measured attenuation values of winter defoliated trees are used to represent these signal losses. Since the loss through the tree in full summer leaf is a power function, the total loss of signal through a tree with foliage

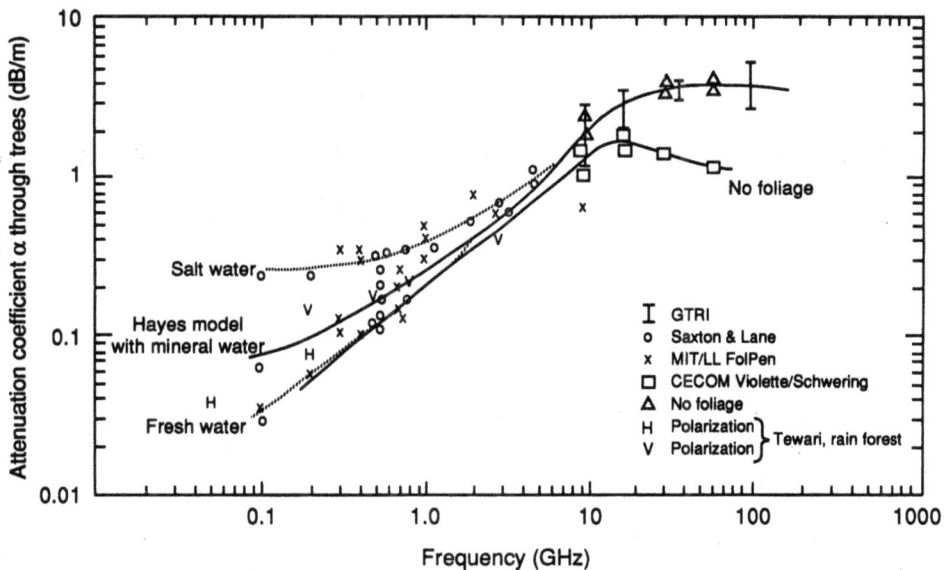

Figure 2.19 Measured and calculated attenuation coefficient for trees versus frequency. (Adapted from Hayes [56].)

is the sum of the winter measured values plus these calculated values. Figure 2.19 shows both measured and calculated values of attenuation coefficients for tree foliage [56].

Measured data have shown no effects of polarization on attenuation for frequencies between 1 GHz and 100 GHz. For frequencies between 20 MHz and 500 MHz, there can be an appreciable difference in attenuation coefficients for horizontal polarization compared to vertical polarization. At the lower frequencies, the loss for vertical polarization is twice that for horizontal polarization. Calculations based on the model generated using forward scattering from leaves, tree leaf density, the complex dielectric constant for water, and measurements made during winter months (no foliage) agree extremely well with measurements made during summer (full foliage) for a wide frequency band from 100 MHz to 100 GHz.

2.6.4 Particulate, Smoke, and Aerosol Attenuation

Recall from Equation (1.40) that the total attenuation of a signal is the sum of absorption plus scattering effects. Thus, the total attenuation (or extinction) coefficient α is the sum of the attenuation coefficient α_s due to absorption, and the attenuation coefficient as due to scattering. Assuming Rayleigh scattering ($\pi D/\lambda \ll 1$), where D is the particle diameter, the attenuation coefficient α_a due to absorption, in dB/m, is given by the expression

$$\alpha_a = 8.186 \frac{\text{Im}(-K)M}{\lambda \rho} \qquad (2.77)$$

and the attenuation coefficient α_s due to scattering, in dB/m, is given by the expression

$$\alpha_s = 169.1 \frac{MD^3K^2}{\lambda^4 \rho} \qquad (2.78)$$

where

M = mass of particles, in g/m^3
ρ = density of particles, in g/cm^3
λ = radar wavelength, in cm
D = particle diameter, in cm [57]

The effects of battlefield obscurants on the performance of millimeter-wave systems have been the subject of several measurement programs. These obscurants' attenuation characteristics were measured during Smoke Week II at frequencies of

35, 94, and 140 GHz [17]. Of these obscurants, the only one that had a significant effect on millimeter-wave transmission was high explosive dust generated by munitions. This attenuation was significant for only a few seconds after the explosion, while large ejecta were still airborne.

These results can be interpreted in terms of particle size. The HC, WP, RP, oil fog, and smoke all have characteristic sizes of a few microns or smaller. Sizes of the aluminum or carbon flakes were not given, but were likely to have been a few micrometers in size. The dust produced by vehicular traffic will be about 50 μm in size; the dust produced by the HE munitions explosions will have particle sizes well in excess of 100 μm. The particles that are much smaller than the wavelength of the incident radiation will have very low efficiencies for attenuation or backscatter. Particles that are larger than 100 μm will have an effect on transmission, but the effect will be short lived due to the extremely rapid settling of these particles. The signal decay data suggest that particle lifetimes are about 20 sec. Other tests during the Misers Bluff experiment for radars at a number of frequencies between 0.4 and 95.9 GHz also showed that explosion-generated dust affected the amplitude and phase shift of radar signals, but the effects also lasted for only a few seconds after the explosion.

Based on the sizes of particles in soils, it was expected that the particles lofted by the explosions would be in the range of 100 to 400 μm, a value consistent with the particle settling times observed. Mie calculations of attenuation efficiencies (attenuation (dB/m) per gram of material per cubic meter) at 35 and 95 GHz for particles with refractive indices characteristic of soil particles are shown in Table 2.10 below [20]. These results show the increase in attenuation efficiency for the 300-μm particles for the 95-GHz frequency. Increases in efficiency at 35 GHz will occur for somewhat larger particles. These Mie calculations indicate that backscatter efficiencies will be several orders of magnitude less than attenuation efficiencies for the 50-μm particles and roughly comparable for the 300-μm particles. Moist soil material will have higher attenuation at both wavelengths, with comparable backscatter values.

It should be noted that although the smoke from combustion was not effective as an obscurant, Ebersole and Vaglio-Laurin [58] discussed measurements that show

Table 2.10
Attenuation Efficiencies α_{eff} for Dry Soil Particles (Adapted from Perry [20])

	$a = 50\ \mu m$	$a = 300\ \mu m$
35 GHz	7.3×10^{-6}	9.3×10^{-6}
95 GHz	2.0×10^{-5}	1.2×10^{-4}

intensity fluctuations of 3 dB in transmitted signals due to scintillation effects caused by the hot gases and water vapor emitted during the combustion process. Such effects were important only in the immediate vicinity of the fires.

These results show that most common battlefield obscurants will not affect millimeter-wave propagation unless the size of the obscuring particles is greater than 100 μm. This occurs because for spherical or near spherical particles, the greatest attenuation efficiency is seen when the particle diameter is approximately equal to the wavelength of incident electromagnetic radiation. As the wavelength increases, the size parameter of the particles relative to wavelength decreases, and the attenuation efficiency decreases, eventually falling off as $1/\lambda^4$ for dielectric particles, as $1/\lambda$ for absorbing particles, and as $1/\lambda^2$ when the particles are conductors that can be described by a free electronic model [59].

Particles that are larger than 100 μm have large settling velocities and cannot remain suspended for long durations; consequently, they are not suitable for use as millimeter-wave obscurants. The need for better broadband obscurants has led to the development of tailored particles with the desired multispectral attenuation characteristics.

The theory of these multispectral obscurants has been developed by several authors, notably Pedersen et al. [60], Daum, [61], and Swinford [62]. Daum discusses the interaction of electromagnetic radiation with matter in terms of harmonic oscillators. Nonconducting particles have tightly bound oscillators, which can absorb energy only at certain frequencies. Scattering will take place at other wavelengths. For spherical or other compact dielectric particles, if the particle size is much smaller than the incident wavelength, the particles will have little effect on the electromagnetic waves, and attenuation efficiencies are small. As the size of the particle approaches the wavelength of the incident waves, the interaction and efficiencies are greatest.

Conducting particles, by contrast, have loosely bound electrons that may be represented as oscillators with many closely spaced frequencies, which can result in broadband absorption. Highly conducting materials behave as free oscillators which relax by re-emitting the absorbed energy at the incident frequency; there is very little absorption, and the attenuation is due to scattering rather than absorption. Lower conductivity materials, such as iron and graphite, have a loss term that causes damping of the induced currents, absorption of the incident energy, and joule heating of the material.

The absorption also depends on the magnitude of the induced dipole moment, which should be maximized for large absorption. This will occur for fibers in which the length is much greater than the cross section radius, with the region of maximum absorption extending to larger wavelengths as the length to diameter (L/d) ratio is increased. The penetration or skin depth in the perpendicular direction will be small for conducting materials. At a 1-mm wavelength, a particle with a conductivity of 10^5 $(\Omega\text{-m})^{-1}$ will have a skin depth of 29 μm; a conductivity of 10^6, a skin depth of 2.9 μm; and a conductivity of 10^7, a skin depth of 0.3 μm. Since there is no increase in absorption for any thickness larger than the penetration depth, particles do not need to be thicker than the penetration depth. The greatest efficiencies will be seen for conducting fiber particles with small radii and with large L/d values; this geometry will also minimize the particle volume.

If the material conductivity, the length, and the fiber radius could all be varied, then, in theory, attenuating particles could be designed for any effect. Figure 2.20 shows the effect of variation in L/d values for a fixed conductivity particle of diameter 0.024 μm [61]. For these particles the attenuation at millimeter wavelengths is primarily due to absorption, and the attenuation in the plateau region between 0.1 and 1.0 mm will be proportional to the conductivity. Calculations show that theoretical attenuation efficiencies of more than 20 dB (m^2/g) are possible. Recall that the units of attenuation efficiency are in attenuation (m^{-1}) per gram of material per cubic meter (g/m^3).

Not all of the possible fiber attenuation characteristics can be achieved. Conductivities will decrease as the fiber thickness approaches the mean free path of the electrons in the conducting material. Generation methods for the fibers may lead to different conductivities from pure bulk material. Perhaps most importantly, there are practical limitations on the fibers that can be generated and disseminated. Highest efficiencies achieved in practice appear to be in the range of 1 to 10 m^2/g.

Two specific sets of calculations for broadband tailored obscurants from Pedersen are shown in Figures 2.21 and 2.22 [63]. The results of these calculations are

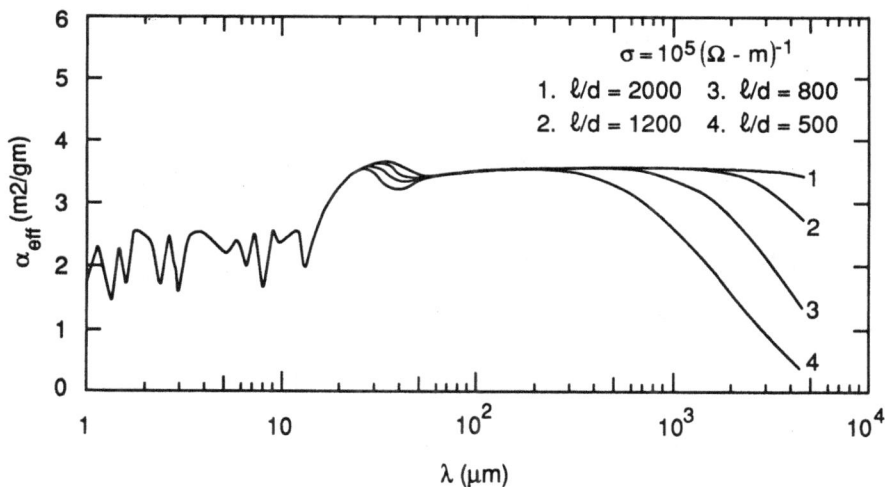

Figure 2.20 Extinction efficiency versus wavelength for a range of L/d ratios. (Adapted from Daum [61].)

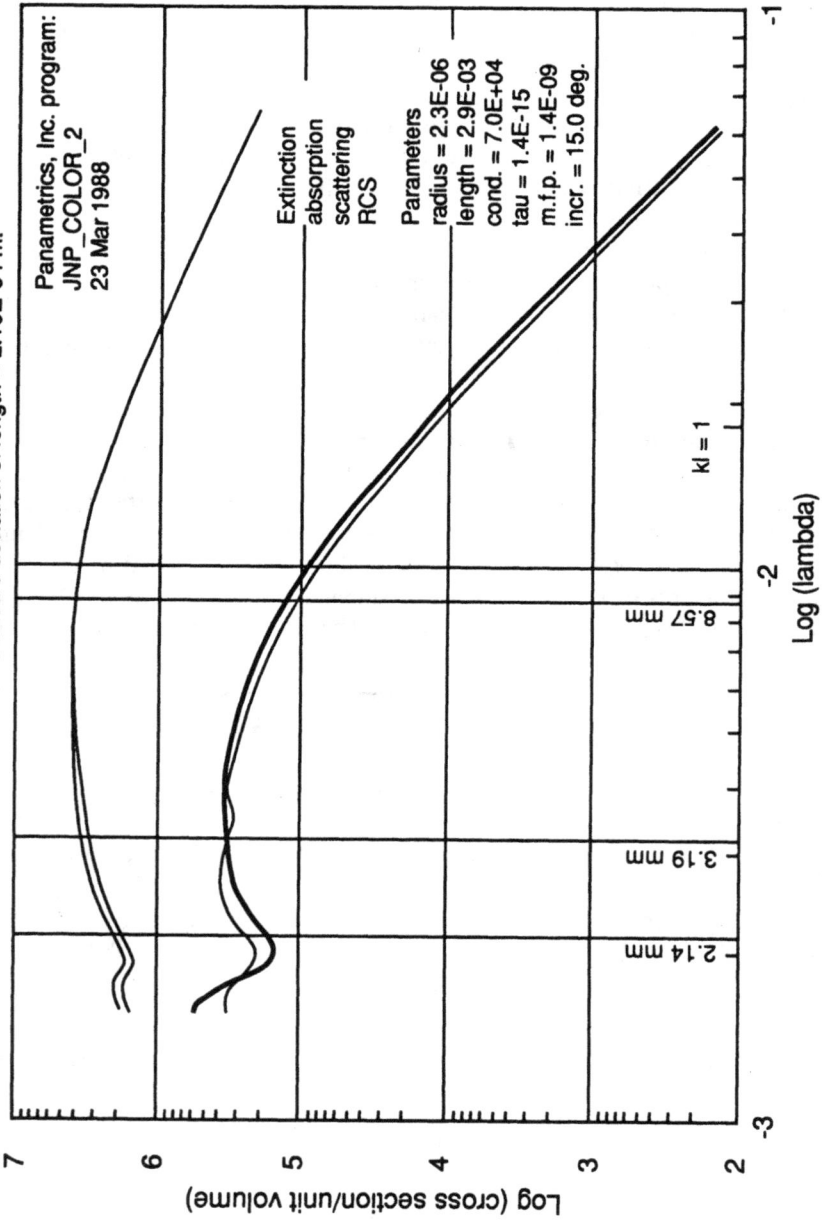

Figure 2.21 Ratio of particle cross section to particle volume versus wavelength. (Adapted from Pederson [63].)

Plots of extinction and scattering cross sections versus wavelength.
3-D data. MKS units.

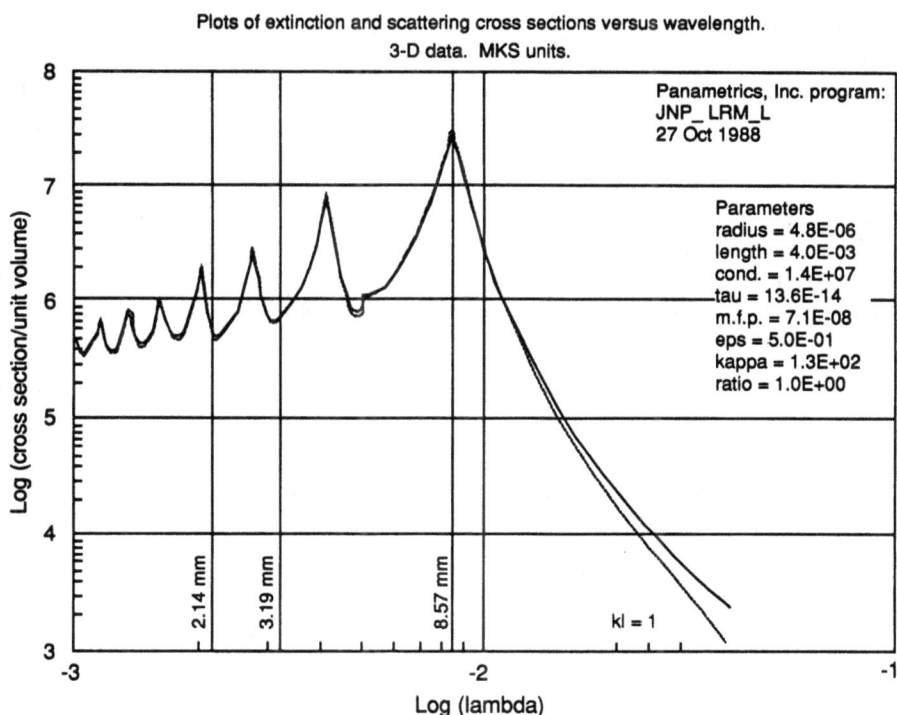

Figure 2.22 Ratio of particle extinction and scattering cross sections to particle volume versus wavelength. (Adapted from Pederson [63].)

expressed as logarithms to base 10 of ε values (ratios of cross section to particle volume). The cross section to particle volume ratios can be converted to the appropriate efficiencies by dividing by particle density.

Figure 2.21 shows the modeling results for fibers with radius equal to 2.3 μm, length of 2.9 mm, and conductivity of 7×10^4 $(\Omega$-m$)^{-1}$. The attenuation is due primarily to absorption, as expected from the level of conductivity. The backscatter or RCS efficiency is more than an order of magnitude less than the attenuation efficiency at 35 GHz. Figure 2.22, by contrast, shows a case in which the attenuation is dominated by scattering due to the higher conductivity of the material. The attenuation cross sections have peak values that are higher than the absorbing case but that fall off much more quickly with increasing wavelength. Backscatter and attenuation efficiencies are both equal to approximately 10 m^2/g at the wavelengths of interest.

Measurements to characterize the attenuation efficiency for particles similar to those modeled in Figure 2.21 have been reported by Bruce, et al. [64]. They found that measured attenuation efficiencies were similar to those predicted on the basis of concentration, microphysical measurements, and an attenuation model.

Sophisticated computer simulations of scattering from conductive fibers have been developed by Pedersen et al. [65], which are very useful in predicting the absorption, scattering, and extinction cross sections of various fibers. Assume a homogeneous obscurant cloud of fiber particles. The cloud can be completely represented by the variables of Table 2.11.

The theory of Pedersen et al. provides estimates of ε_b and ε_e, and chemical analysis of the particle provides ρ. With these values, the following relationships can be generated for scattering cross section and particle efficiency:

$$\alpha_b = \frac{\varepsilon_b}{\rho} \tag{2.79}$$

and

$$\alpha_{\text{eff}} = \frac{\varepsilon_{\text{eff}}}{\rho} \tag{2.80}$$

Table 2.11
Obscurant Cloud Attenuation and Backscattering Variables
(Adapted from Pederson, Waterman, and Pederson [65])

Variable	Units	Definition
l	m	Length of cloud
w	m	Width of cloud
h	m	Height of cloud
V	m^3	Volume of cloud $= l\,w\,h$
f	g/m^3	Molecular density of individual particles
ρ	g/m^3	Mass density of an individual particle
c	g/m^3	Concentration of cloud, as a whole
M	g	Mass of material in cloud
ε_b	m^2/m^3	Backscatter cross section per particle
ε_e	m^2/m^3	Extinction (attenuation) cross section per particle
α_b	m^2/g	Backscatter efficiency
α_{eff}	m^2/g	Extinction (attenuation) efficiency
T	Np	Transmittance through cloud
A	dB	Total one-way attenuation through cloud (of length l)
α	dB/m	Attenuation coefficient of cloud
η	m^2/m^3	Backscatter coefficient of cloud

Now, the transmittance through the cloud is given by Beer's law:

$$T = \exp(\alpha_{\text{eff}}) \exp(cl) \tag{2.81}$$

from which the attenuation coefficient α can be derived. Define L, the one-way loss in dB, to be

$$L = -10 \log T = 4.3 \alpha_{\text{eff}} cl \tag{2.82}$$

so

$$\alpha = L/l = 4.3 \alpha_{\text{eff}} c \tag{2.83}$$

and

$$\alpha = 4.3 \frac{\varepsilon_{\text{eff}} M}{\rho V} \tag{2.84}$$

Similarly, the following expression defines the backscatter coefficient η:

$$\eta = \alpha_b c = \frac{\varepsilon_b M}{\rho V} \tag{2.85}$$

Note the similarity between the relationships of equations (2.84) and (2.85). Comparison of the two equations leads to the following relationships, from which several observations may be drawn:

$$\eta(\text{m}^2/\text{m}^3) = 0.233 \, (\varepsilon_b/\varepsilon_{\text{eff}}) \alpha \qquad (\text{dB}/\text{m}) \tag{2.86}$$

$$\phi \, \text{dB}(\text{m}^{-1}) = 10 \log \alpha + 10 \log [0.233(\varepsilon_b/\varepsilon_{\text{eff}})] \tag{2.87}$$

and

$$\alpha \, (\text{dB}/\text{m}) \propto c \tag{2.88}$$

while

$$\eta \, \text{dB}(\text{m}^{-1}) \propto \log c \tag{2.89}$$

- The backscatter η, in *linear* units (m^2/m^3), is directly proportional to the attenuation coefficient α in *logarithmic* units (dB/m).

- For a particular particle (oriented randomly), both η and α are simply functions of concentration c.
- The attenuation coefficient is much more sensitive to changes in concentration than is the backscatter coefficient.

REFERENCES

[1]   E.F. Knott, J.F. Shaeffer, and M.T. Tuley, *Radar Cross Section,* Artech House, Inc., Norwood, Massachusetts, 1985, pp. 89–90.

[2]   M. Abramowitz and I. Stegun, eds., *Handbook of Mathematical Functions,* National Bureau of Standards Applied Mathematics Series, No. 55, December 1972, p. 358.

[3]   R.N. Trebits, "Radar Cross Section," Chapter 1 in *Techniques of Radar Reflectivity Measurement,* N.C. Currie, ed., Artech House, Inc., Norwood, Massachusetts, 1984, p. 34.

[4]   E.F. Knot, J.F. Shaeffer, and M.T. Tuley, *Radar Cross Section,* Artech House, Inc., Norwood, Massachusetts, 1985, pp. 177–180.

[5]   H.A. Corriher, Jr., et al., "Clutter Characteristics and Effects, Chapter XXI in *Principles of Modern Radar,* Georgia Tech Short Course Notes, 1980, p. 332.

[6]   W.E.K. Middleton, *Vision Through the Atmosphere,* University of Toronto Press, Toronto, 1963.

[7]   R.G. Eldridge, "Haze and Fog Aerosol Distributions," *Journal of Atmospheric Science,* Vol. 23, September 1966, pp. 601–613.

[8]   D. Atlas, "Advances in Radar Meteorology," Advances in Geophysics, Volume 10, Academic Press, New York, 1964, pp. 317–478.

[9]   H. Sauvageot, *Radar Meteorology,* Artech House, Inc., Norwood, Massachusetts, 1992, p. 312.

[10]   J.S. Marshall and W.M. Palmer, "The Distribution of Raindrops With Size," *Journal of Meteorology,* Vol. 5, August 1948, pp. 165–166.

[11]   J.O. Laws and D.A. Parsons, "The Relation of Raindrop Size to Intensity," *Transactions of the American Geophysical Union,* Vol. 24, 1943, pp. 452–460.

[12]   N.C. Currie, F.B. Dyer, and R.D. Hayes, "Analysis of Radar Rain Return at Frequencies of 9.375, 35, 70, and 95 GHz," Technical Report No. 2, Contract DAAA 25–73–0256, Georgia Institute of Technology, February 1975.

[13]   H. Sauvageot, *Radar Meteorology,* Artech House, Inc., Norwood, Massachusetts, 1992, p. 81.

[14]   K.L.S. Gunn and J.S. Marshall, "The Distribution With Size of Aggregate Snowflakes," *Journal of Meteorology,* Vol. 15, 1958, pp. 452–466.

[15]   R.S. Sekhon and R.C. Srivastava, "Snow Size Spectra and Radar Reflectivity," *Journal of Atmospheric Science,* Vol. 27, pp. 299–307.

[16]   P.M. Austin, "Radar Measurements of the Distribution of Precipitation in New England Storms," *Proceedings of the 10th Weather Radar Conference,* 1963, pp. 247–254.

[17]   J.E. Knox, "Millimeter Wave Propagation Measurements at Smoke Week II," Memorandum Report ARBRL-MR-03002, U.S. Army Research and Development Command, Ballistic Research Laboratory, Arberdeen Proving Ground, Maryland, March 1980.

[18]   R.D. Hayes, "Radar Scattering and Absorption by Sand," RDH, Inc., unpublished paper, February, 1989.

[19]   C.R. Thomas and J.E. Cockayne, "MISER Bluff II Cloud Sampling Program," Report DNA 001–78-C-0217, Science Applications, Inc., for the Defense Nuclear Agency, December 1979.

[20]   B. Perry, personal communications, Marietta, Georgia, February 1992.

[21]   M.W. Long, *Radar Reflectivity of Land and Sea,* 2nd ed., Artech House, Inc., Norwood, Massachusetts, 1983, pp. 37–39.

[22]   N.C. Currie, "Clutter Characteristics and Effects," Chapter 10 in *Principles of Modern Radar,"* J.L. Eaves and E.K. Reedy, eds., Van Nostrand-Reinhold Company, New York, pp. 281–342.

[23]   R.L. Cosgriff, W.H. Peake, and R.C. Taylor, "Terrain Scattering Properties for Sensor System Design," Terrain Handbook II, Engineering Experiment Station Bulletin, Vol. 29, No. 3, The Ohio State University, 1960.

[24]   R.D. Hayes et al., "Study of Polarization Characteristics of Radar Targets," Final Technical Report, Contract No. DA-36–039-SC-64713, Georgia Institute of Technology, Atlanta, Georgia, October 1958.

[25]   R.D. Hayes, "95 GHz Pulsed Radar Returns from Trees," *IEEE EASCON 1979 Conference Record,* October 1979.

[26]   M.W. Long, *Radar Reflectivity of Land and Sea,* Artech House, Inc., Norwood, Massachusetts, 1984, pp. 189–210.

[27]   M.W. Long, *Radar Reflectivity of Land and Sea,* Artech House, Inc., Norwood, Massachusetts, 1984, pp. 362–370.

[28]   N.C. Currie, F.B. Dyer, and R.D. Hayes, "Radar Land Clutter Measurements at Frequencies of 9.5, 16, 35, and 95 GHz," Technical Report No. 3, Contract DAAA 25–76–0221, Georgia Institute of Technology, Atlanta, Georgia, April 1975, ADA 012709.

[29]   I.S. Gradshteyn and I.M. Ryzhik, *Table of Integrals, Series, and Products,* translated by Scripta Technica, Inc., A. Jeffrey, ed., Academic Press, New York.

[30]   N.C. Currie and S.P. Zehner, "Millimeter Wave Clutter Model Update," *IEE Radar 87 Digest,* London, October 1987.

[31]   W.H. Stiles and F.T. Ulaby, "Microwave Remote Sensing of Snowpacks," University of Kansas Center for Research, Inc., NASA Contractor Report 3263, 1980.

[32]   S. Evans, "Dielectric Properties of Ice and Snow: A Review," *Journal of Glaciology,* Vol. 5, 1965, p. 773.

[33]   A.T. Edgerton, A. Stogryn, and G. Poe, "Microwave Radiometric Investigation of Snowpacks," Aerojet-General Corporation, Final Report No. 1285 R-4, USGS Contract No. 14–08–001–11828, July 1971.

[34]   S. Twomey, H. Jacobowitz, and H.B. Howell, "Matrix Methods for Multiple-Scattering Problems," *Journal of Atmospheric Sciences,* Vol. 23, 1966, pp. 289–296.

[35]   F.T. Ulaby, A.K. Fung, and W.H. Stiles, "Backscatter and Emission of Snow: Literature Review and Recommendations for Future Investigations," RSL Technical Report 369–1, University of Kansas, Center for Research, Inc., for the U.S. Air Force Armament Laboratory, June 1978, pp. 97–101.

[36]   R.A. Bolander and R.W. McMillan, "Atmospheric Effects on Near-Millimeter-Wave Propagation," *Proceedings of the IEEE,* Vol. 73, No. 1, January 1985, pp. 49–60.

[37]   J.C. Wiltse, "Introduction and Overview of Millimeter Waves," Chapter 1 in *Infrared and Millimeter Waves,* K.J. Button and J.C. Wiltse, eds., Volume 4, *Millimeter Systems,* Academic Press, New York, 1981, p. 4.

[38]   J.H. Van Vleck, "The Absorption of Microwaves by Oxygen," *Physical Review,* Volume 71, April 1947, pp. 413–424.

[39]   B.R. Bean, and E.J. Dutton, *Radio Meteorology,* Dover Publications, New York, 1968.

[40]   K.L.S. Gunn and T.W.R. East, "The Microwave Properties of Precipitation Particles," *Quarterly Journal of the Royal Meteorological Society,* Vol. 80, October 1954, pp. 533–545.

[41]   E.A. Altschuler, "A Simple Expression for Estimating Attenuation by Fog at Millimeter Wavelengths," *IEEE Transactions on Antenna and Propagation,* Vol. AP-32, No. 7, July 1984, pp. 757–758.

[42]   N.P. Robinson, "Measurement of the Effect of Rain, Snow, and Fog on 8–6 mm Radar Echoes," *Proceedings of the IEE,* Vol. 203-B, September 1955, pp. 709–714.

[43]   S.M. Kulpa and E.A. Brown, Co-chairmen, "Near-Millimeter Wave Technology Base Study," Harry Diamond Laboratory Report No. HDL-SR-79–8, November 1979, p. 37.

[44]   R.N. Trebits, "MMW Propagation Phenomena," Chapter 4 in *Principles and Applications of Millimeter-Wave Radar,* N.C. Currie and C.E. Brown, ed., Artech House, Inc., Norwood, Massachusetts, 1987, pp. 131–188.

[45]   H. Sauvageot, *Radar Meteorology,* Artech House, Inc., Norwood, Massachusetts, 1992, p. 107.

[46]   R.D. Hayes, Private Communication, RDH, Inc., January 1992.

[47]   T. Oguchi, "Scattering from Hydrometeors: A Survey," *Radio Science,* Vol. 16, No. 5, September-October 1981, pp. 691–730.

[48]   T. Oguchi, "Rain Depolarization Studies at Centimeter and Millimeter Wavelengths: Theory and Measurement," *Journal of Radio Research Laboratories,* Volume 22, No. 109, 1975, pp. 165–211.

[49]   A. Hendry and G.C. McCormick, "Polarization Properties of Precipitation Scattering," pp. 9–20.

[50]   T. Oguchi and Y. Hosoya, "Differential Attenuation and Differential Phase Shift of Radio Waves Due to Rain: Calculations at Microwave and Millimeter Wave Regions," *Journal of Rech. Atmosphere,* Vol. 8, 1974, pp. 121–128.

[51]   Commission of the European Communities, 1978.

[52]   A.J. Bogush, Jr., *Radar and the Atmosphere,* Artech House, Inc., Norwood, Massachusetts, 1989, p. 332.

[53]   A. Nishitsuji and A. Matsumato, "Calculation of Radio Wave Attenuation Due to Snowfall," SHF and EHF Propagation in Snowy Districts, Monograph of the Research Institute of Applied Electricity, Hokkaido University, No. 19, 1971, pp. 63–78.

[54]   E.F. Knott, Private Communication, 1985.

[55]   J.S. Rothacher, F.E. Blow, and S.M. Potts, "Estimating the Quantity of Tree Foliage in Oak Stands in the Tennessee Valley," *Journal of Forestry,* March 1954.

[56]   R.D. Hayes, "Attenuation Through Trees," RDH, Inc., unpublished paper, May 1990.

[57]   R.D. Hayes, "Radar Scattering and Absorption by Sand," RDH, Inc., unpublished paper, February 1989.

[58]   J.F. Ebersole and R. Vaglio-Laurin, "An Assessment of Battlefield-Induced Aerosol Contaminants," Technical Report ARCSL-CR-80068 for U.S. Army Armament Research and Development Command Chemical Systems Laboratory, Aerodyne Research, Bedford, Massachusetts, September 1980.

[59]   J.J. Savage, "Smoke Research at CRDEC," Proceedings of the Smoke/Obscurants Symposium XIV, Volume I, CRDEC-CR-092, Science and Technology Corporation, Hampton, Virginia, pp. 3–6.

[60]   N.E. Pedersen, P.C. Waterman, and J.C. Pedersen, "Electromagnetic Cross Sections of Conductive Fibers: Modified Drude Equations and Dependence of Dielectric Constant on Particle Size," Report AFOSR-TR-88–1019, for U.S. Air Force Office of Scientific Research, Panametrics, Inc., Waltham, Massachusetts, August 1988.

[61]   G.R. Daum, "The Theory of Multispectral Screening," Technical Report BRL-TR-2693, U.S. Army Ballistic Research Laboratory, November 1985, p. 54.

[62]   H.W. Swinford, "Electromagnetic Behavior of Radar Absorbing Chaff," Technical Note 354–43, Naval Weapons Center, China Lake, California, June 1975.

[63]   Pederson, personal communication.

[64]   C.W. Bruce, A.V. Jelinek, R.M. Halonen, and M.J. Stehling, "Millimeter Wavelength Attenuation Efficiencies of Fibrous Aerosols," Proceedings of the Smoke/Obscurants Symposium

XIV, Volume I, CRDEC-CR-092, Science and Technology Corporation, Hampton, Virginia, pp. 119–126.

[65] N.E. Pedersen, P.C. Waterman, and J.C. Pedersen, "Electromagnetic Cross Sections of Conductive Fibers: Modified Drude Equations and Dependence of Dielectric Constant on Particle Size," Report AFOSR-TR-88–1019, for U.S. Air Force Office of Scientific Research, Panametrics, Inc., Waltham, Massachusetts, August 1988.

Chapter 3
Clutter Characteristics

3.1 WHAT ARE THE CHARACTERISTICS OF CLUTTER?

As we discussed in Chapter 1, clutter is the return from a real object or group of objects that interferes with the signal from the target of interest to the radar. An additional consideration for this book is the fact that millimeter wavelengths are the focus for consideration. Because of Mie scattering and other nonlinear scattering effects unique to the MMW region, as discussed in Chapter 2, it is not possible to simply scale X-band data upwards to 95 GHz and expect to achieve a realistic approximation of reality!

For analytical purposes, clutter is conventionally divided into two primary categories: volume (atmospheric) clutter and surface (land and sea) clutter, because of the differing geometries and the varied characteristics of the types of clutter. These characteristics are discussed separately in this chapter.

3.1.1 Atmospheric Clutter

One way to describe atmospheric clutter is to examine how it appears to an operator on a radar display. Atmospheric clutter has two major effects on a radar signal: it attenuates the signal traveling through the atmosphere to and from a target of interest (through absorption and scattering in directions away from the radar target line of sight), and it reflects energy in a pseudorandom fashion back towards the radar. If this reflected, or backscattered, energy falls within the beamwidth of the radar and is located near the target within the radar range resolution, then the clutter signal will interfere with the target signal. The attenuation effect lowers the reflected signal level from the target, and the intracell scattering effect raises the apparent noise level of the receiver. Both effects lower the detectability of the target.

When heavy precipitation occurs or manmade chaff is dispensed, the reflectivity of the clutter can actually saturate the receiver over a certain range, thus ren-

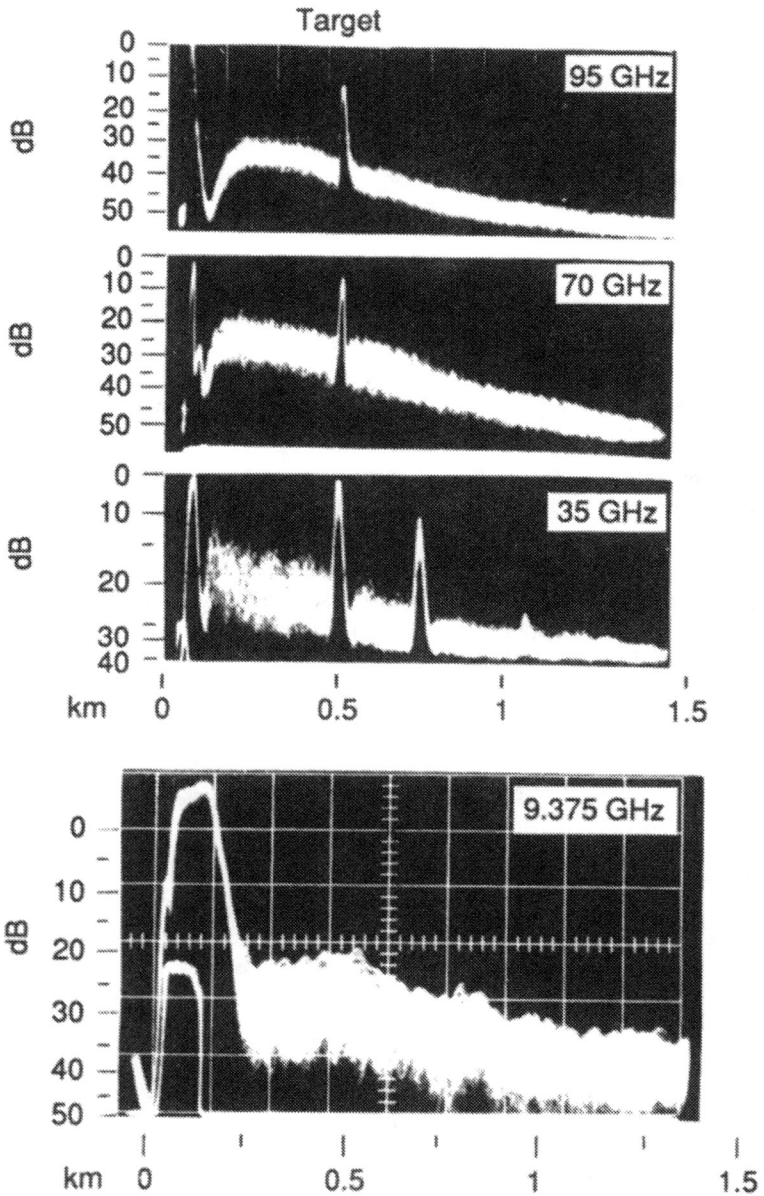

Figure 3.1 A-scope photograph of the return from rain at four frequencies. (From Downs [1].)

dering a target completely undetectable to the radar. Figure 3.1 shows A-scope displays (amplitude versus range) of the return from a heavy rain at 9, 35, 70, and 95 GHz [1]. Two trihedral reflectors are shown at a range of 0.5 and 0.75 km. The rapid trail-off of the rain return at 35 GHz and above is due to the heavy attenuation at these higher MMW frequencies.

Atmospheric reflections appear as a "blob" on an area display, such as a plan position indicator (PPI), which displays range and azimuth in a polar format. Figure 3.2 shows a PPI display of heavy rain returns at X-band. Light rain may not show up on a display; rather, it may require that the radar receiver gain be increased as the target signal decreases, and the radar sensitivity may decrease as the apparent

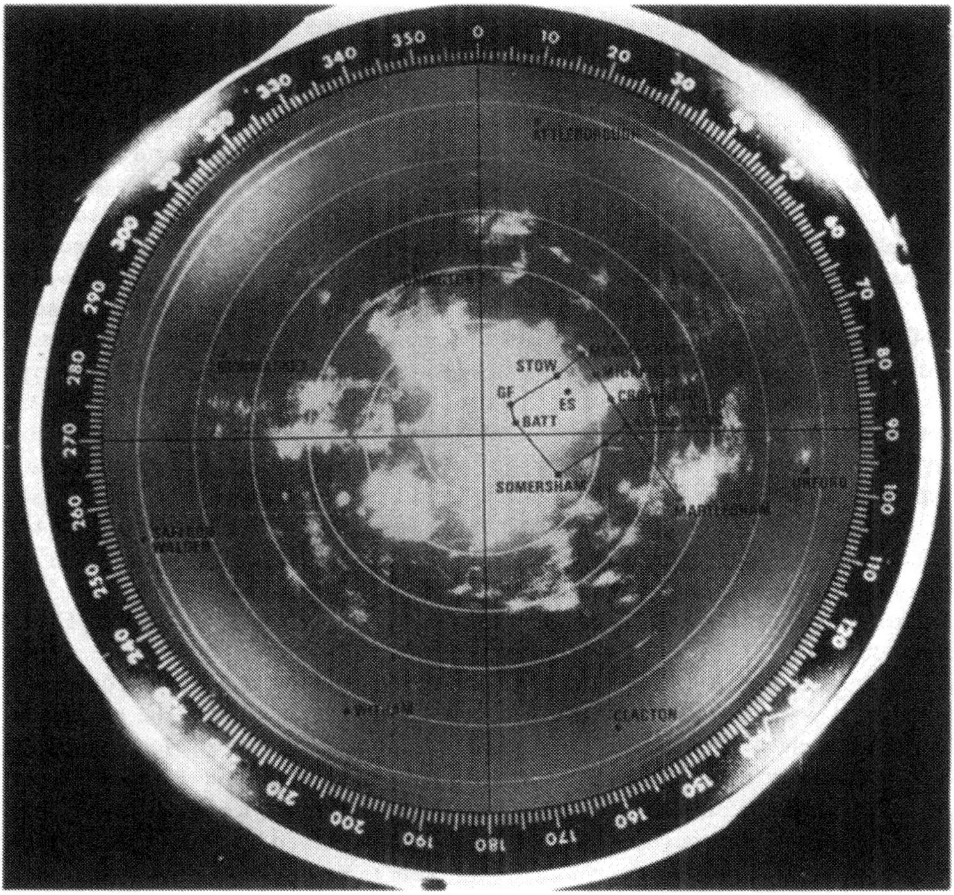

Figure 3.2 PPI radar display of heavy rain showers in Great Britain. (From Turner [2].)

noise floor increases. Other types of atmospheric returns may have similar effects (snow, chaff, and heavy dust/smoke). Since the volume illuminated by the radar as defined by the beamwidth increases as the square of the range to a target, and the signal from a target or a single clutter scatterer decreases as the fourth power of the range, the range dependence of atmospheric clutter (ignoring attenuation) is $1/R^2$, whereas the range dependence of a target is $1/R^4$. Thus, there is a maximum range for which the target is lost in the clutter, which is much less than the noise-limited range, even for a large radar cross section (RCS) target.

3.1.2 Surface Clutter

Surface clutter is generally divided into land clutter and sea clutter because of the different characteristics of each type. Land clutter consists of a patchwork quilt of clutter returns that may be uniform over a limited area, but quite different from patch to patch. Embedded in these uniform patches are point-like clutter returns similar to (and may consist of) manmade targets. Sea clutter is relatively uniform over a relatively large area, but it varies with the radar line of sight to the wind-wave direction. Sea return varies with time as the wind speed and direction change. Embedded in Rayleigh-like sea returns are target-like returns called *spikes*. Spikes have a number of causes, including birds, white caps, and concave "facets" in the sea waves.

Figure 3.3 shows a synthetic aperture radar (SAR) image of a rural scene at 35 GHz, which illustrates the patchwork appearance of land clutter. For this reason, clutter measurements are generally confined to carefully measuring a particular type of relatively uniform clutter, such as grass, crops, and trees, and combining the various types of uniform clutter in a simulation. One of the key problems for this approach, however, is the "blending" [3] of the edges of the uniform patches in a realistic manner, since many false alarms from actual clutter occur at clutter transitions, such as at tree lines, which create havoc with constant false alarm rate (CFAR) detectors. Many such edges can be seen in Figure 3.3. Also, many point targets are visible in which a return from a relatively small area is significantly larger than the surrounding returns. Such areas may result in false alarms from simple CFAR processors.

Figure 3.4 gives a high-resolution sea return B-scope image (horizontal axis — azimuth, vertical axis — range) for a "fully developed sea" [4]. The upwind direction of highest sea return is clearly a 45° diagonal from left to right. Although the sea clutter at close range has saturated the radar receiver, specific wave fronts are visible in the middle of the display, high returns from the top of the waves, and low returns from the troughs. The wave fronts appear curved because of the B-scope distortion. At first glance, sea clutter would seem to be less of a problem than land clutter because of its lower reflectivity values and its relative uniformity. Unfortunately, sea targets of interest, such as small boats, submarine periscopes, and swim-

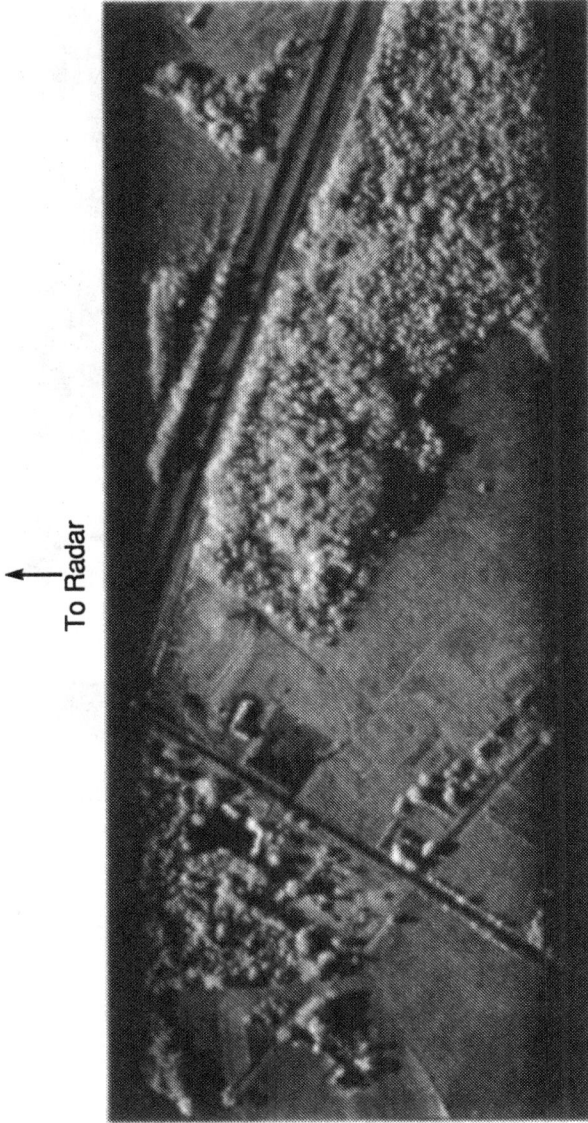

Figure 3.3 Synthetic aperture radar image at 35 GHz of land terrain showing varied "textures" of clutter. (Photo courtesy of Massachusetts Institute of Technology Lincoln Laboratory (MIT/LL).)

Figure 3.4 B-scope radar display of the Atlantic Ocean near Boca Raton, Florida, at X-band. (From Long [4].)

mer delivery vehicles (SDV), are much smaller, both physically and in RCS, than typical land targets such as tanks, armored personnel carriers (APC), and trucks. The lone exceptions are large naval vessels and ships.

3.2 SPECIFIC CHARACTERISTICS

In this section, specific data available in the literature on millimeter wave clutter will be summarized. The characteristics will be presented in terms of the amplitude, temporal, and spatial parameters discussed in Chapter 2. Where available, models currently used to describe such clutter will be included.

3.2.1 Volume Clutter

3.2.1.1 Overview

Volume clutter affects radar detection as discussed above through attenuation and backscatter. Hydrometeors (rain, snow, clouds, and hail) are the primary volume clutter encountered, although dust, dirt, explosive ejecta, and chaff are also occasionally seen. As discussed in Chapter 2, hydrometeor reflections tend to occur in the Mie or optical scattering regions for millimeter wave radars, so that wide variation in the data and poor predictability tend to be the rule. In addition, because of limited radar power and higher attenuation in the medium, measured data on atmospheric clutter become scarce as the frequency rises.

3.2.1.2 Rain Clutter Data

Amplitude Characteristics

Average Values. The definitive experiment to measure rain backscatter at millimeter waves was performed by the Ballistic Research Laboratory (BRL) in 1973, in which rain backscatter and attenuation were measured at 10, 35, 70, and 95 GHz over an instrumented range for which several sets of tipping buckets were instrumented to provide simultaneous rain rate measurement. Georgia Tech (GT) participated in the experiment by recording data on magnetic tape for later computer analysis of spectra and correlation properties, as well as amplitude distribution, since BRL used A-scope photographs to determine amplitudes. Significant variation in the data were observed from measurement to measurement, as indicated in Figure 3.5, which compares the 95-GHz rain backscatter data obtained by BRL and GT with least square fits (LSF) to the data. The averaging time was approximately 0.1 sec for the BRL data and 30 sec for the GT data. These fluctuations in radar return were due to time varying drop size distributions and the subsequent effect on raindrop RCS. In spite of this wide variation, the LSF curves given in Figure 3.6 show remarkable similarities between the BRL and GT data. The only major difference is that the BRL data show 95-GHz data slightly higher than the 70-GHz data, while the GT data indicate that the 70-GHz data are higher in magnitude than the 95. The 10- and 35-GHz data LSF lines are virtually identical. Figure 3.7 gives the Georgia Tech data for circular polarization. The curves are virtually identical in shape to the linear (vertical) except for being 6 to 10 dB lower in value. BRL reported reductions in rain reflectivity of up to 20 dB for circular polarization as opposed to vertical, but did not present LSF curves.

From a number of measurements made by radar meteorologists employing microwave radars during the 1950s, summarized by Skolnik [7], Nathanson [8], Barton

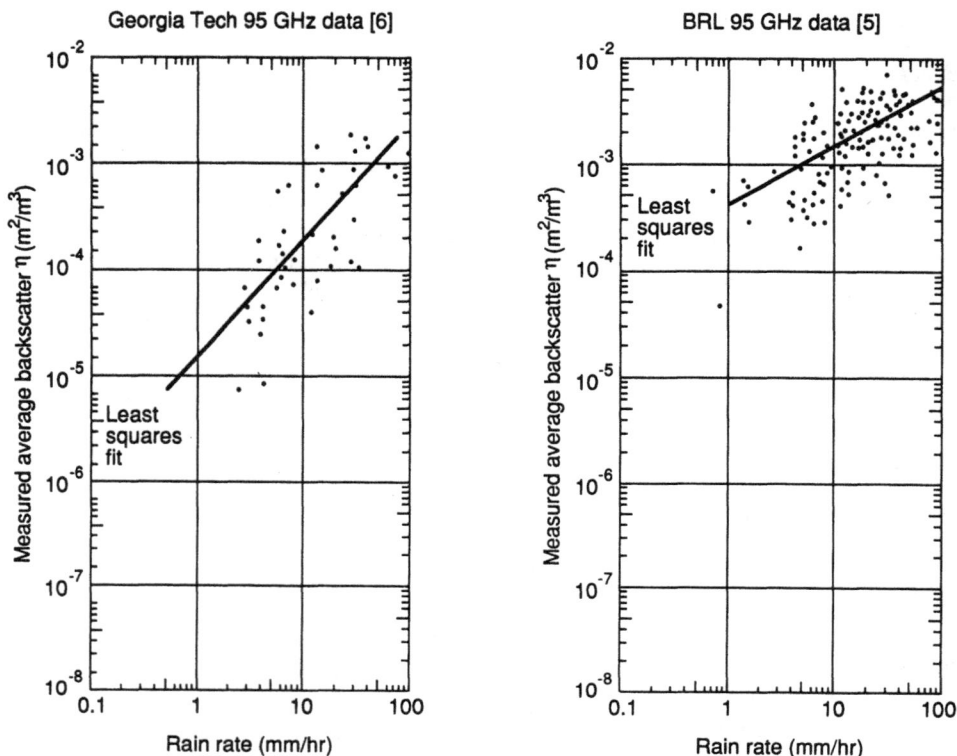

Figure 3.5 Comparison of Ballistic Research Laboratory and Georgia Tech rain data with LSF curves at 95 GHz. (From Currie, Dyer, and Hayes [5] and Richard and Kammerer [6].)

[9], and Beam [10], values of the scattering coefficient Z for drop size distributions of rains were developed, where Z is of the form

$$Z = \sum D^6 N(D_i) \tag{3.1}$$

and where D is the drop diameter and $N(D_i)$ is the drop size distribution function.

Table 3.1 gives the measured values for Z at microwave frequencies for various types of storms and rainfall rates R.

If we use the value $Z = 200\, R^{1.6}$ in Equation (2.11) for the volume backscatter coefficient, and the value of $|K|^2 = 0.93$ for water, then the volume backscatter coefficient η is expressed by

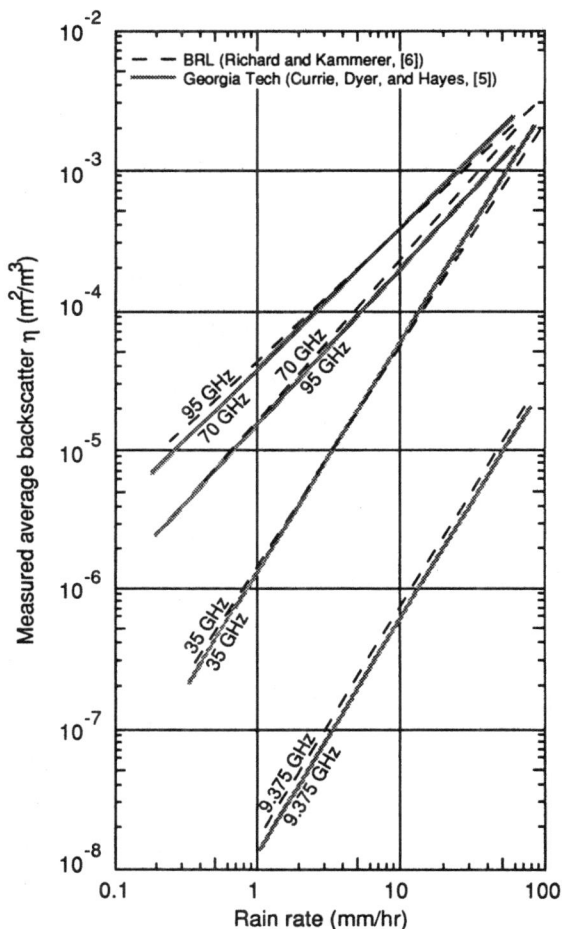

Figure 3.6 Comparison of Ballistic Research Laboratory and Georgia Tech LSF curves for rain backscatter, vertical polarization.

$$\eta = \frac{\pi^5 f^4}{(3 \times 10^8)^4} (0.93) \frac{200 R^{1.6}}{10^{-18}} \tag{3.2}$$

$$\eta = 7 \times 10^{-12} R^{1.6} f^4, \ (\text{m}^2/\text{m}^3) \tag{3.3}$$

where f is the frequency in GHz.

Comparing this equation to rain backscatter data reported by Georgia Tech [5], the equation seems to represent the peak values as a function of the rain rate R.

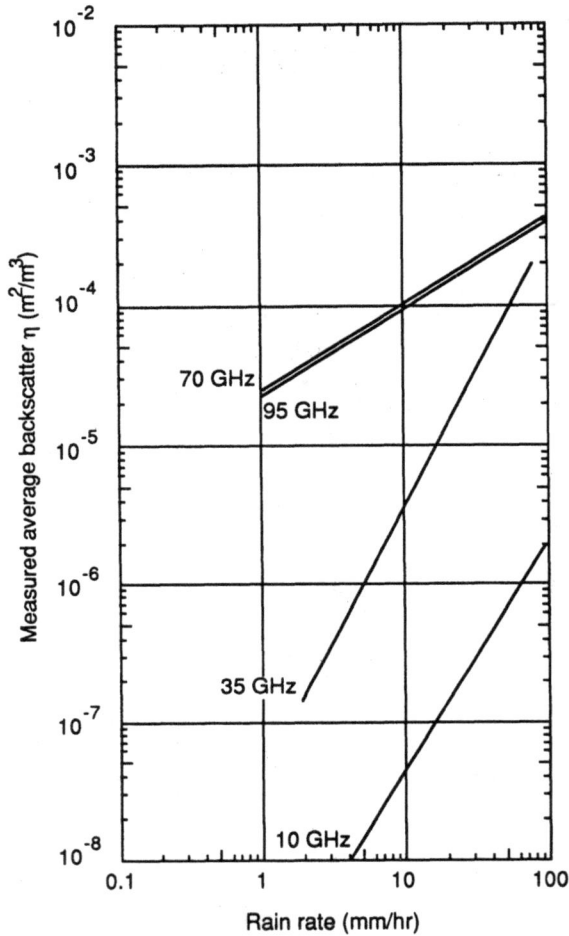

Figure 3.7 Georgia Tech rain data LSF curves for rain backscatter, circular polarization. (From Currie, Dyer, and Hayes [5].)

Figures 3.8 and 3.9 show these equations plotted with the Georgia Tech data from [5] and an LSF curve. Figure 3.10 compares an equation for 95 GHz that also represents the peak value for 95-GHz data. These data have a different exponential dependence than the 10-GHz and 35-GHz data.

Models. Tables 3.1 and 3.2 give the equations and typical values for the LSF values shown in Figures 3.6 and 3.7. Table 3.2 includes the equation for previous microwave measurements plotted in Figures 3.8 and 3.9, which appear to represent peak

Table 3.1
Measured Values of Z at Microwave Frequencies

Type of Rain	$Z \ (mm^6/m^3)$
Stratiform	$200 \ R^{1.6}$
	$214 \ R^{1.39}$
	$286 \ R^{1.43}$
	$311 \ R^{1.44}$
Orographic	$31 \ R^{1.71}$
Thunderstorm	$486 \ R^{1.37}$

where R is the rain rate in mm/hr.

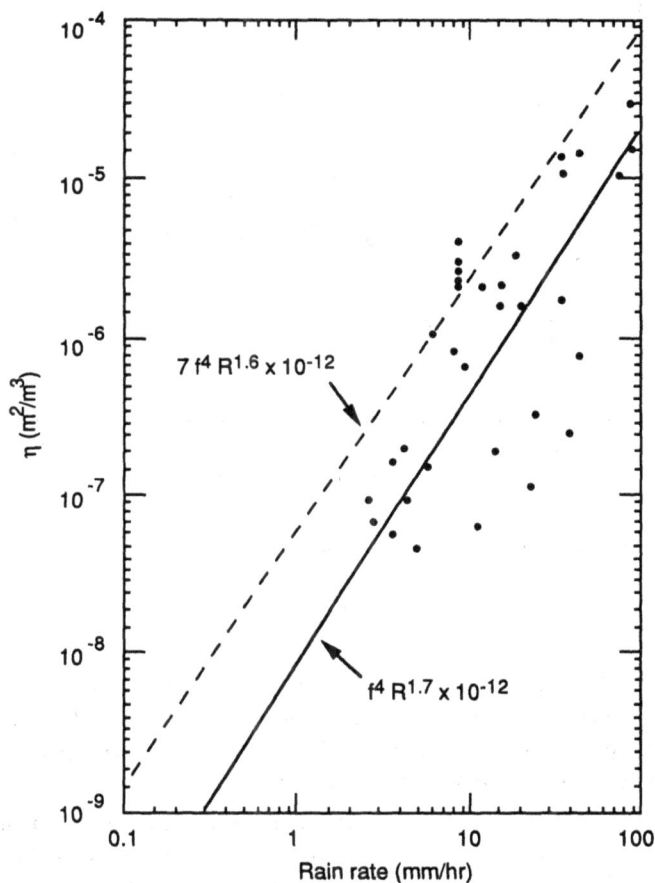

Figure 3.8 Comparison of equation based on microwave measurements to GT rain backscatter data and LSF at 10 GHz.

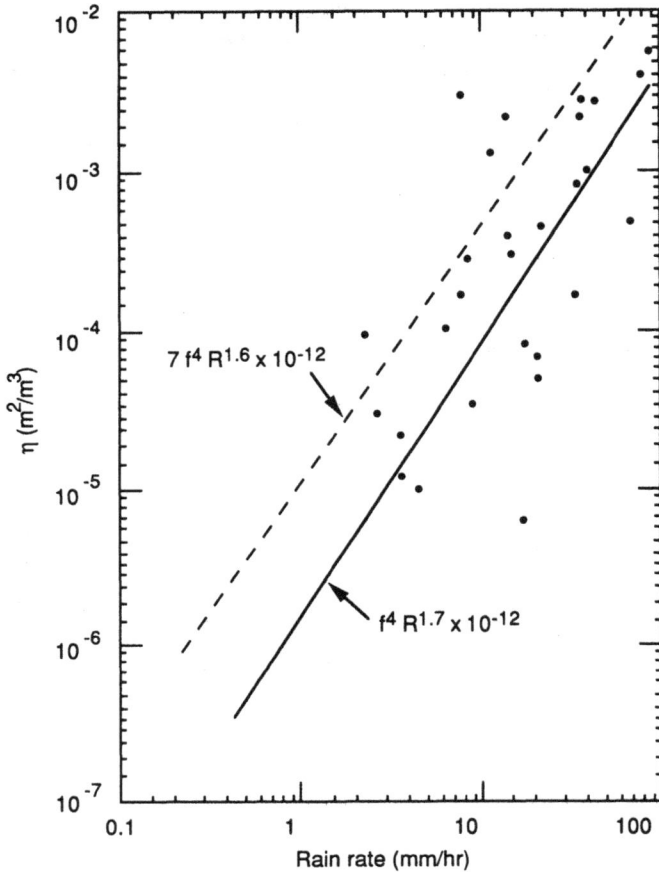

Figure 3.9 Comparison of equation based on microwave measurements to GT rain backscatter data and LSF at 35 GHz.

values, as well as the equation for peak values for 95 GHz from Figure 3.10. These equations can be used to calculate values of η for rain backscatter from 10 to 35, 70, and near 95 GHz. Table 3.3 gives the BRL LSF equations. The reader can use the LSF to the BRL or GT data, or the equations.

Amplitude Statistics. Traditionally, the return from rain has been considered to be Rayleigh distributed. (See Ch. 1 for a definition of the Rayleigh distribution.) Figure 3.11 illustrates such a Rayleigh distribution at X-band. However, at millimeter-wave frequencies the returns have been measured, and the distributions appear to better

Table 3.2

Model for GT LSF to Rain Backscatter Data (Vertical Polarization)

$$\eta = AR^B$$

R = rain rate; A, B as given below

Frequency	A	B	Source
10	1.1×10^{-8}	1.7	LSF to data
10	$7f^4 \times 10^{-12}$	1.6	Peak calculation
35	$7f^4 \times 10^{-12}$	1.7	LSF to data
35	1.5×10^{-6}	1.6	Peak calculation
70	3.3×10^{-5}	1.1	LSF to data
95	1.46×10^{-5}	1.1	LSF to data
95	$1.25f^4 \times 10^{-12}$	1.1	Peak calculation

f is the frequency in GHz.

Table 3.3

Model for BRL LSF to Rain Backscatter Data

$$\eta = AR^B$$

R = rain rate; A, B as given below

Frequency	A	B	Source
10	3.98×10^{-8}	1.5	LSF to data
35	8.498×10^{-5}	1.05	LSF to data
70	5.55×10^{-4}	0.58	LSF to data
95	13.42×10^{-4}	0.57	LSF to data

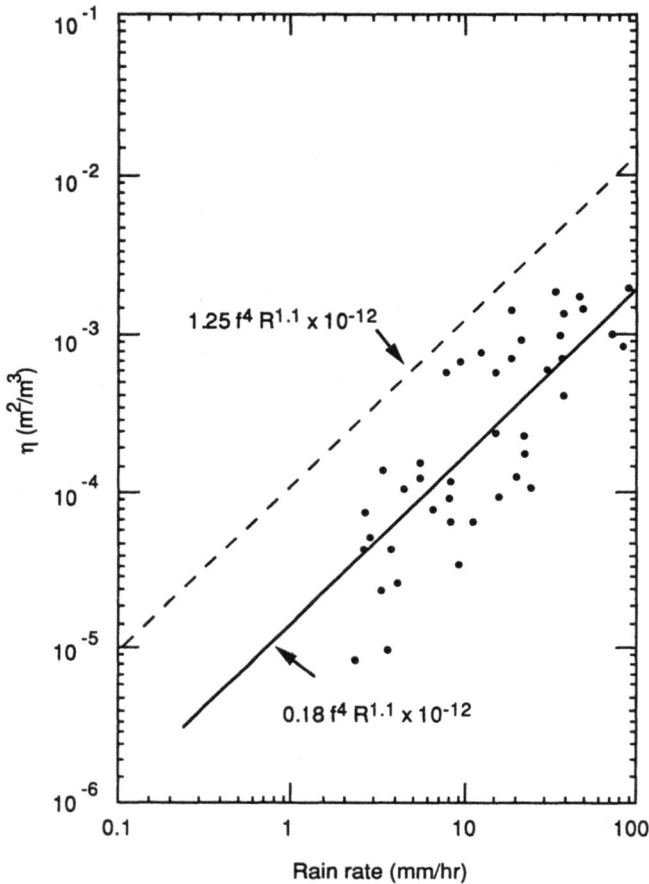

Figure 3.10 Comparison of equation representing peak return to GT rain backscatter data and LSF at 95 GHz.

approximated by lognormal distributions, as shown in Figures 3.12 and 3.13. For the cumulative distributions shown, lognormally distributed data would appear as a straight line, which closely approximates the actual data shown. The apparent non-Rayleigh statistics could be due to unusual drop size distributions generated by the thunderstorms, or due to the use of logarithmic receivers in the experiment. Figures 3.14 and 3.15 give measured standard deviations for the rain returns at 35 and 95 GHz. The standard deviations typically range between 2.5 and 6 dB. For a Rayleigh distribution, a standard deviation of approximately 3.7 dB would be expected.

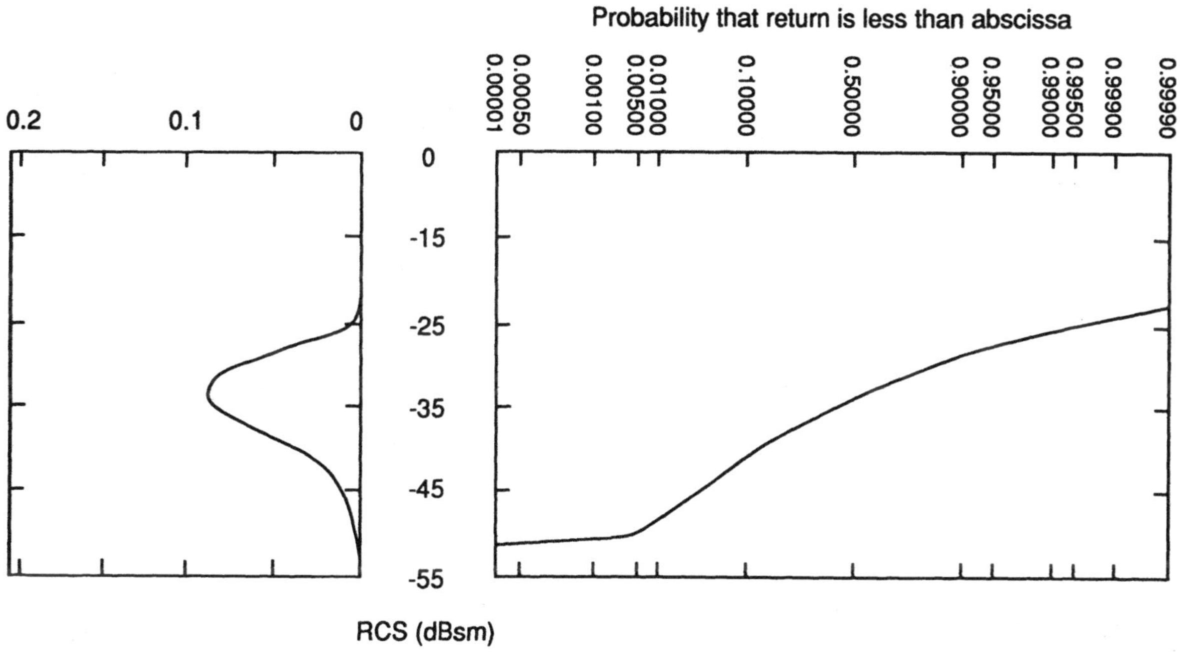

Figure 3.11 Density and cumulative distributions for rain backscatter at 10 GHz. (From Currie, Dyer, and Hayes [5].)

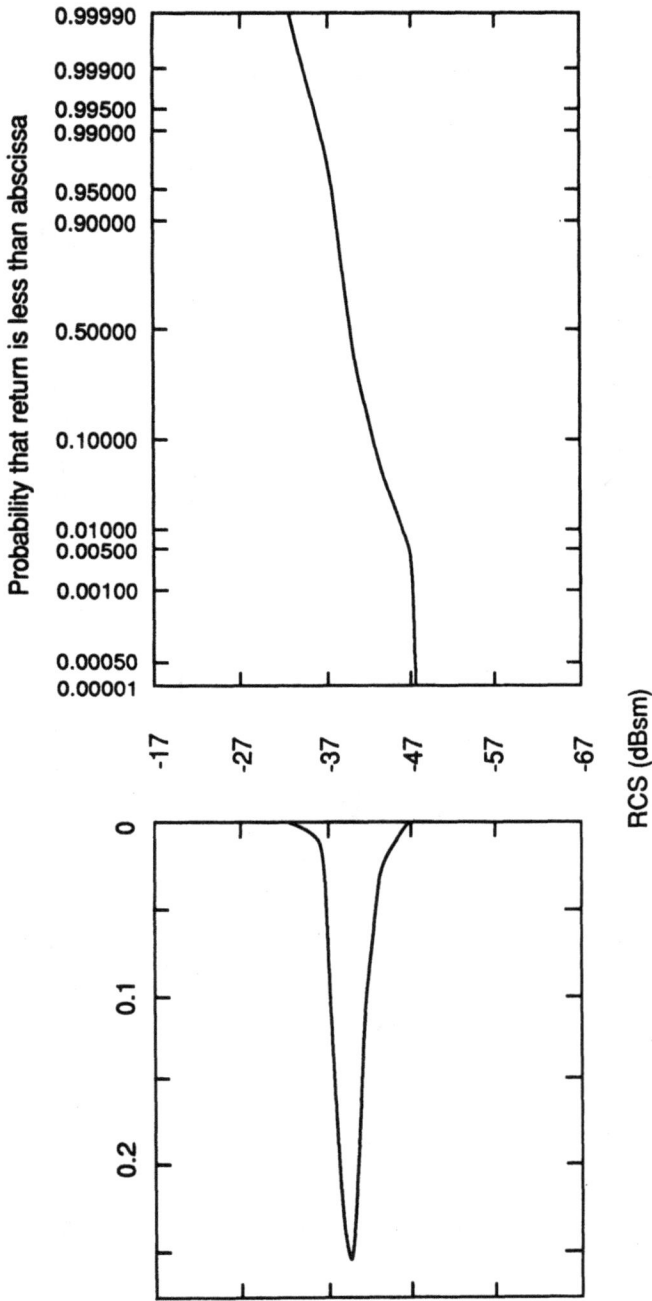

Figure 3.12 Density and cumulative distributions for rain backscatter at 35 GHz. (From Currie, Dyer, and Hayes [5].)

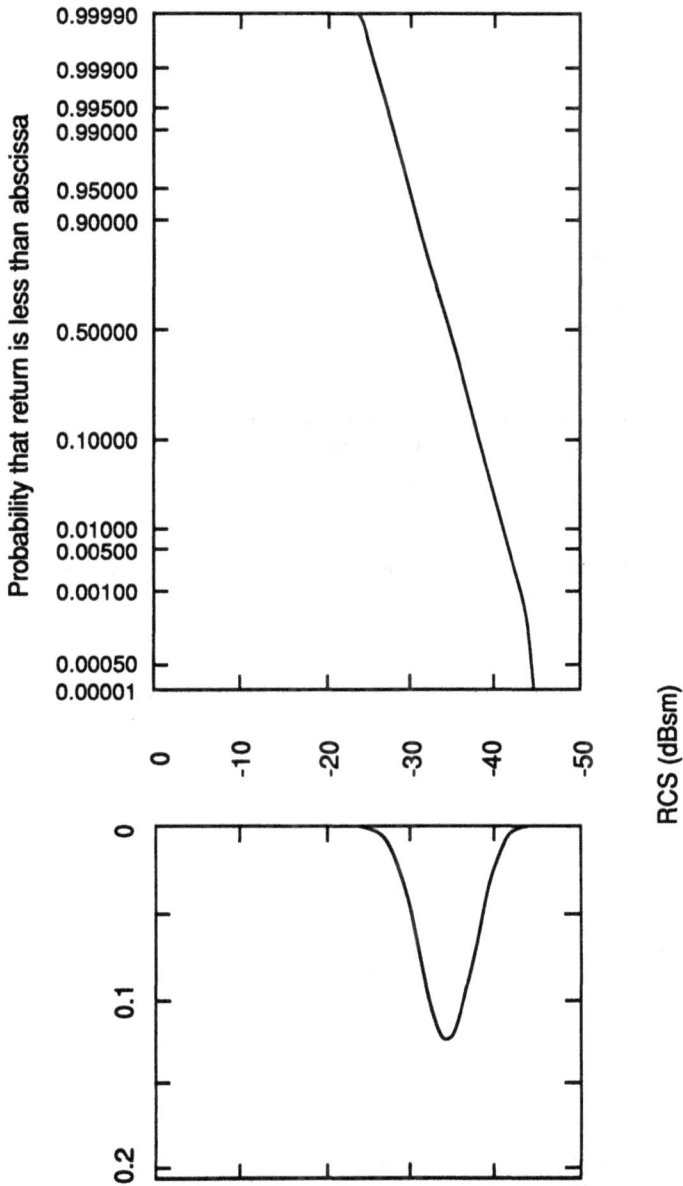

Figure 3.13 Density and cumulative distributions for rain backscatter at 95 GHz. (From Currie, Dyer, and Hayes [5].)

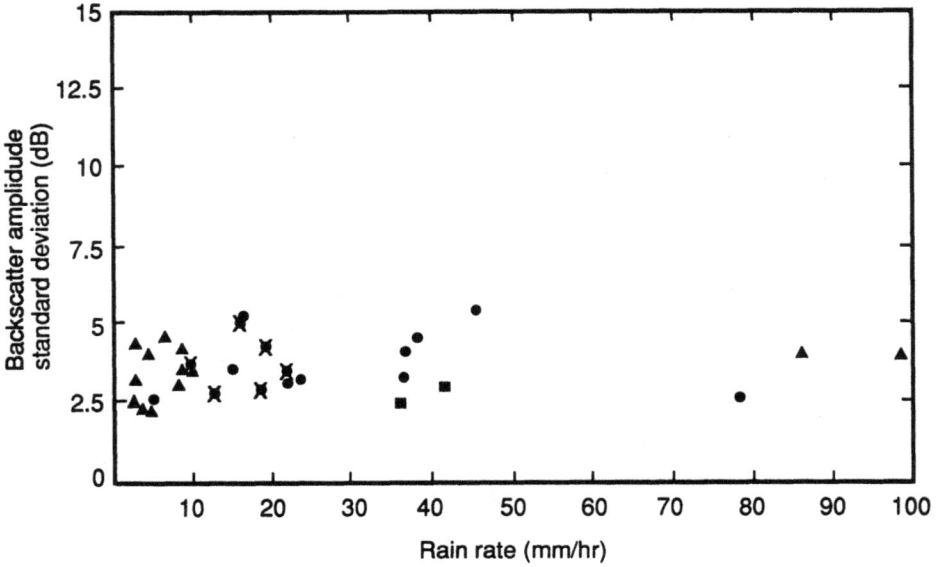

Figure 3.14 Standard deviation versus rain rate for 35-GHz rain data. (From Currie, Dyer, and Hay [5].)

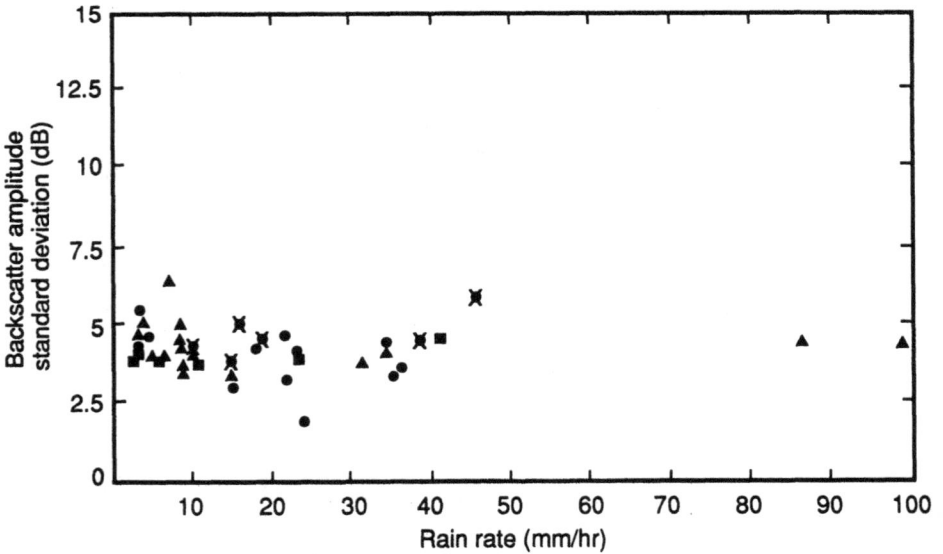

Figure 3.15 Density and cumulative distributions for rain backscatter at 95 GHz. (From Currie, Dye and Hayes [5].)

Temporal Variations

Frequency Spectra. Since falling raindrops are often blown by the wind, a spectral broadening of the return signal is created by the phase changes of the signal scattered from each drop as the round trip path length varies. Such spectral broadening interferes with the ability of a Doppler or moving-target indicator (MTI) radar to detect a moving target. Traditionally, a Gaussian spectrum has been assumed for clutter spectra based on randomly sized scatterers moving at random velocities, as discussed in Chapter 2, but over the years non-Gaussian spectra have been measured by several experimenters [5, 11] for noncoherent clutter measurements. Coherent measurements of L-band and X-band tree clutter were measured by Billingsly and Larabee [12], and the data were analyzed in terms of equivalent Doppler velocity. They found that the energy at levels below −60 dB decay approximately exponentially with increasing Doppler velocity at both L- and X-bands.

However, experimenters have found that the power spectral density functions of noncoherent clutter spectra appear to follow a Lorentzian form:

$$W(f) = \frac{B}{1 + (f/f_c)^x} \qquad (3.4)$$

where

B = magnitude near zero frequency
f_c = corner (half-power) frequency, in Hz
x = 3 for S-, C-, X- and K_u-bands
x = 2 for K_a- and M-bands

Critics of such forms for noncoherent spectra assume that nonlinearities in the measurement systems resulted in the altered forms, but Lorentizian or combinations of Gaussian and Lorentzian spectral shapes have been seen over many years with measurement results from both linear and nonlinear systems for rain, trees, and sea clutter [5, 11, 13].

Analysis by Georgia Tech of the spectra of rain returns from the BRL experiment described above yielded spectral densities of a Lorentzian form, as illustrated in Figures 3.16 to 3.19. Admittedly, the logarithmic receiver used by BRL to achieve wide dynamic range may have affected the spectral shapes, but, to the authors' knowledge, no other MMW spectral data for rain are presently available for comparison. Studies performed with tree backscatter have indicated that the nonlinear logarithmic process tends to broaden the spectral distribution [14]. However, the existing evidence for Lorentzian noncoherent spectra for other types of clutter measured using linear receivers leads us to believe these measured data until new data prove otherwise. Perhaps a reader can provide us with some coherent MMW spectral data on rain for analysis.

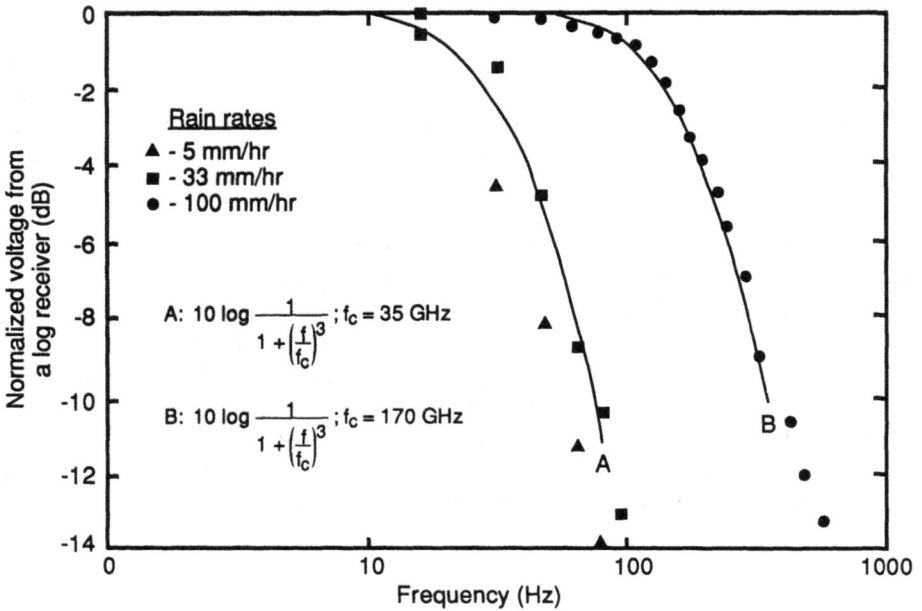

Figure 3.16 Averaged frequency spectra and fitted curves for 10-GHz rain data. (From Currie, Dyer and Hayes [5].)

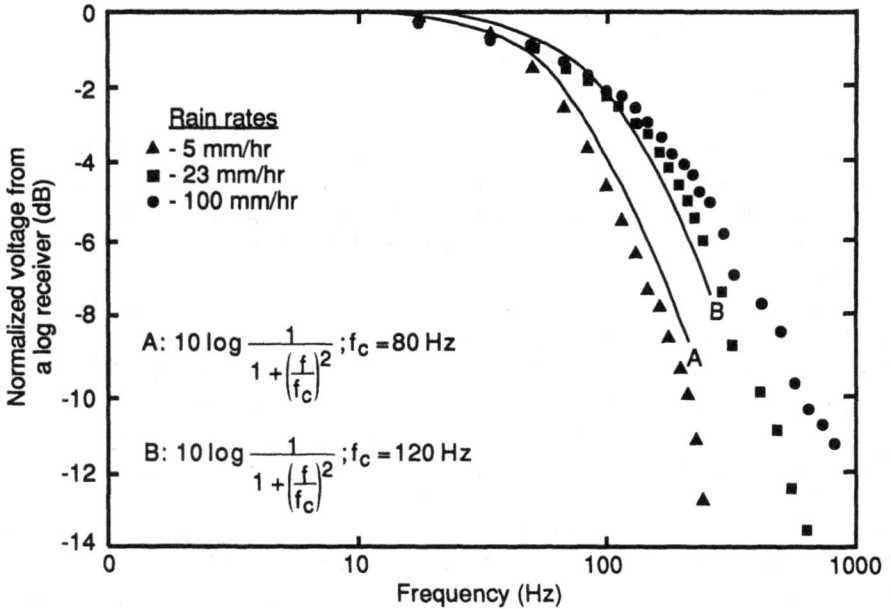

Figure 3.17 Averaged frequency spectra and fitted curves for 35-GHz rain data. (Adapted from Currie, Dyer, and Hayes [5].)

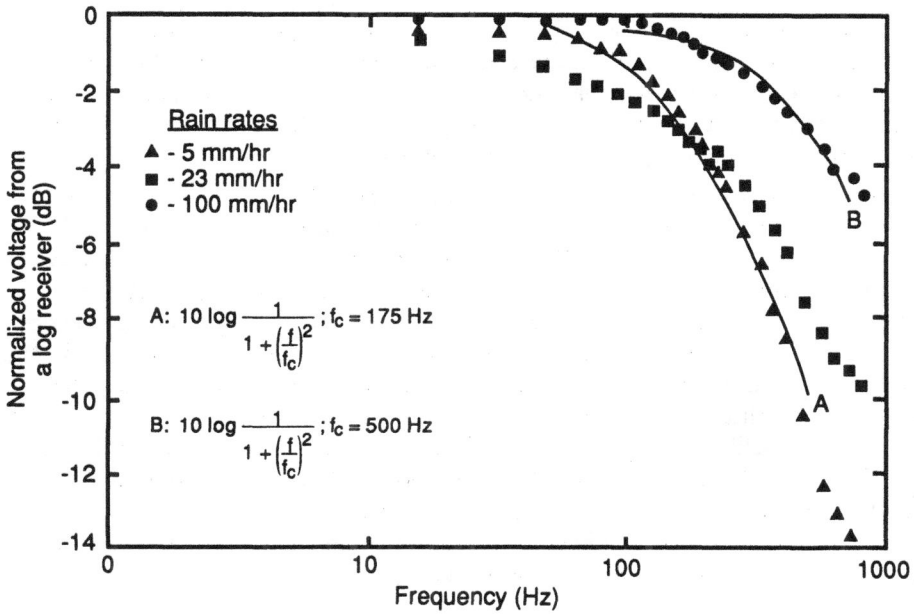

Figure 3.18 Averaged frequency spectra and fitted curves for 70-GHz rain data. (Adapted from Currie, Dyer, and Hayes [5].)

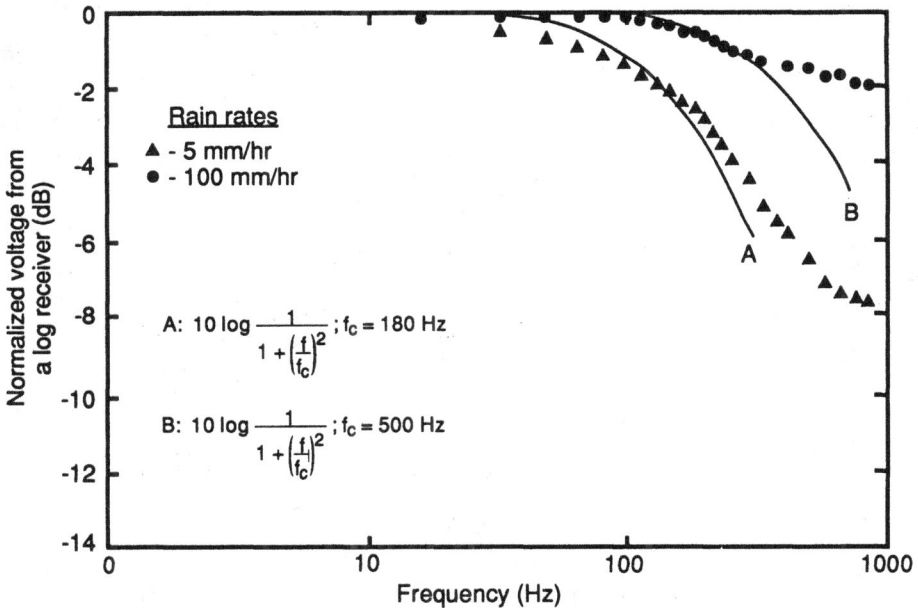

Figure 3.19 Averaged frequency spectra and fitted curves for 95-GHz rain data. (Adapted from Currie, Dyer, and Hayes [5].)

The Georgia Tech data indicate that the corner frequency f_c increases with increasing radar frequency and rain rate. The possible physical cause of the Lorentzian spectra is subject to debate. One physical explanation is the effect of turbulence in thunderstorms (since the data analyzed were from thunderstorms), which might give rise to higher frequency components than normally predicted.

Autocorrelation Functions. Another way of describing the frequency spectrum of the return from clutter is the autocorrelation function, since the power spectral density and autocorrelation function are Fourier transform pairs (see Sec. 1.5.2). However, the autocorrelation function is often used in detection analysis in a different manner than that for frequency spectra. The autocorrelation function gives an indication of the time required for a given signal to decorrelate (i.e., change significantly). Thus, if the decorrelation time (the time for the autocorrelation function to decay to some percent of its initial value, by convention, $1/e$) is known, then the maximum sample rate that will yield uncorrelated (noise-like) samples can be determined. If the sample rate (i.e., the prf) exceeds this value, then no additional integration gain can be achieved by averaging the additional samples so obtained. The time for the autocorrelation function to decay to $1/e$ or 0.37 of its initial value is the decorrelation time. Figure 3.20 summarizes the measured decorrelation times (and the correspond-

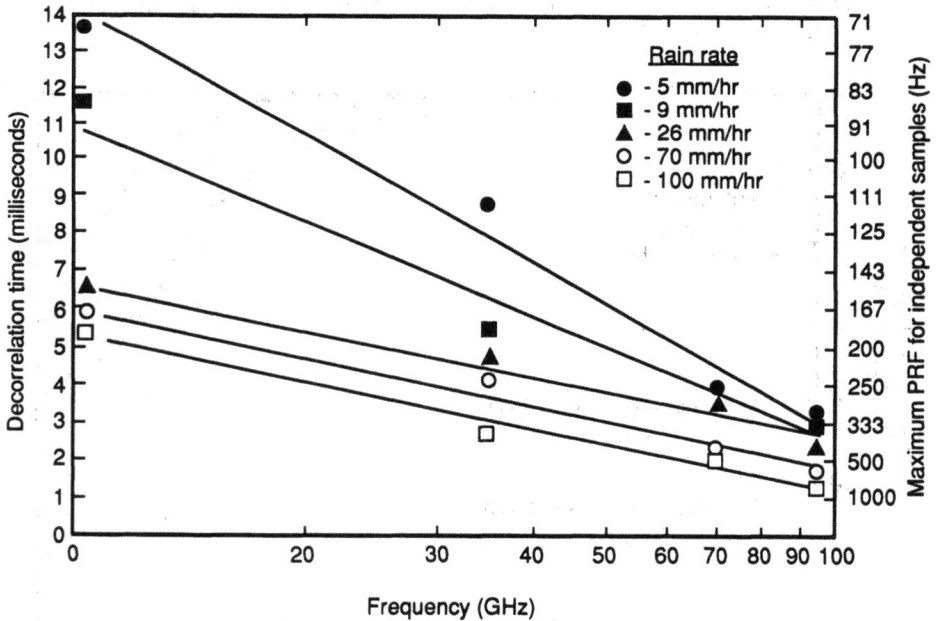

Figure 3.20 Measured decorrelation times for rain versus rain rate and frequency. (Adapted from Currie, Dyer, and Hayes [5].)

ing maximum independent sample rate) for rain as a function of frequency and rain rate. As can be seen, the decorrelation time decreases with increasing radar frequency and rain rate. The decorrelation time is a function of the wind shear and the beamwidth, as discussed by Nathanson [15] for air search radars looking up into the atmosphere. For these data the beamwidths were constant for each frequency band (1°), and a number of data samples from different storms were averaged in order to remove wind shear effects. Thus, these data should represent the internal turbulence of the rain cell and give the relative decorrelation times as a function of radar frequency.

Models. Table 3.4 summarizes the spectral characteristics $W(f)$ for rain backscatter from 10 to 95 GHz.

Table 3.4
Spectral Characteristics of Rain

$$W(f) = \frac{A}{1 + \left(\dfrac{f}{f_c}\right)^N}$$

Frequency (GHz)	Rain Rate (mm/hr)	f_c	N
10	5	35	3
	100	170	3
35	5	80	2
	100	120	2
70	5	170	2
	100	500	2
95	5	200	2
	100	500	2

Table 3.5 summarizes the decorrelation time τ for rain backscatter from 10 to 95 GHz. The table also provides the maximum radar prf that can be used to obtain independent clutter samples.

Attenuation

Measured Data. As discussed in Chapter 2, the generally accepted equation for attenuation coefficient α through aerosols in the atmosphere is of the form

$$\alpha = AR^B \tag{3.5}$$

Table 3.5
Decorrelation Time τ for Rain Backscatter

Frequency (GHz)	Rain Rate (mm/hr)	τ(ms)	Maximum prf
10	5	13.7	73
	9	11.5	87
	26	6.6	151
	70	5.9	169
	100	5.5	182
35	5	8.9	112
	9	5.8	172
	26	5.1	196
	70	4.2	238
	100	3	333
70	5	4.5	222
	9	—	—
	26	3.7	270
	70	2.8	357
	100	2.2	454
95	5	4	250
	9	3.3	300
	26	2.7	370
	70	2	500
	100	1.4	714

where A and B are constants that depend on the drop size distribution and dielectric constant of the aerosols. Measurements of rain attenuation at millimeter wavelengths have been performed by a number of experimentalists, and empirical determination of A and B has been accomplished. Mink [16] measured the rain attenuation at 35 GHz over a short path, with the data resulting in Figure 3.21. The LSF curve to the data is

$$\alpha = 0.539R^{0.811} \tag{3.6}$$

Mink measured an average value for A of 0.539 and an average value for B of 0.811, which were based on measurements performed over several days.

As a part of the rain backscatter experiment performed by BRL [5], the attenuation due to rain was measured at 70 and 95 GHz. These measurements are summarized in Figures 3.22 and 3.23. LSF curves fit to the data yielded

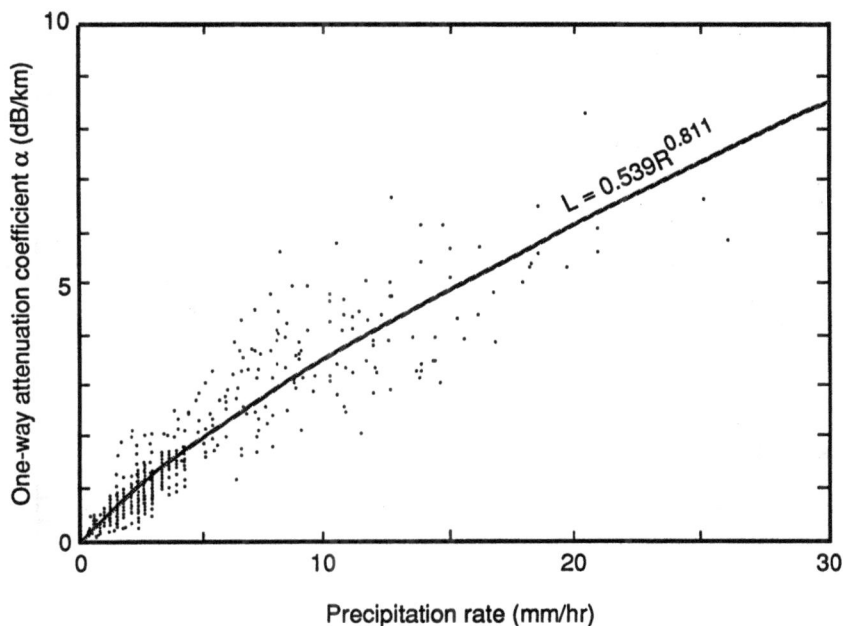

Figure 3.21 Measured attenuation coefficient α for 35-GHz rain data. (Adapted from Mink [16].)

$$\alpha = 0.24R^{1.05} \text{ (70 GHz)} \tag{3.7}$$

$$\alpha = 0.52R^{0.86} \text{ (95 GHz)} \tag{3.8}$$

Keitzer, Sneider, and de Hann of TNO in the Netherlands [17] measured rain attenuation at 94 GHz and compared the results with theoretical calculations using various drop size distributions, including Joss et al. (drizzle), Marshall and Palmer, Laws and Parsons, Joss et al. (thunderstorm), and Joss et al. (widespread), as shown in Figure 3.24. The Marshal and Palmer and Laws and Parsons distributions seemed to fit the data best, although all of the distributions fit within the spread of the data.

Nemarich, Wellman, and Lacombe [18] of the U.S. Army Harry Diamond Laboratory (HDL) performed attenuation measurements on rain at 96, 140, and 225 GHz. Figures 3.25 through 3.27 give measured attenuation data at 96, 140, and 225 GHz, respectively, compared to data calculated using an AR^B relationship and the Laws and Parson low and high rain rate distribution and the Marshall-Palmer distribution. The Laws and Parsons low rain rate distribution calculation fit the data

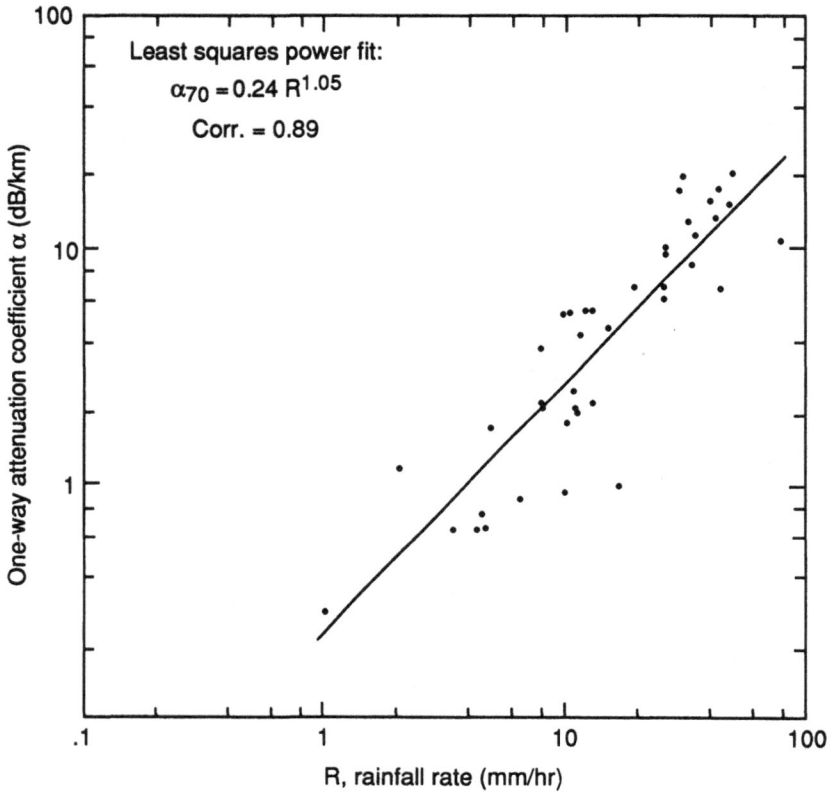

Figure 3.22 Measured attenuation coefficient α for 70-GHz rain data. (Adapted from Richard and Kammerer [6].)

best. At 95 GHz the TNO data appear slightly higher (approximately 2 to 3 dB) than the BRL and HDL data shown in Figures 3.23 and 3.24. Comparing Figures 3.25 through 3.27, we see that the attenuation increases slightly as the frequency increases from 96 to 225 GHz.

V. Furuhama et al. [19] measured the attenuation at 60, 81, 140, and 245 GHz over a horizontal path while measuring the rain rate at both ends of the 1-km path. Figures 3.28 and 3.29 summarize the measurements at each frequency for which rain rates (up to 60 mm/hr) were obtained that were higher than the previously presented data. The trend of slight increase in attenuation coefficient with increasing frequency appears to hold true at higher rain rates as well as lower ones.

Models. Georgia Tech adopted a model for attenuation through rain from 10 through 95 GHz based on an analysis of available theoretical models in the literature of the

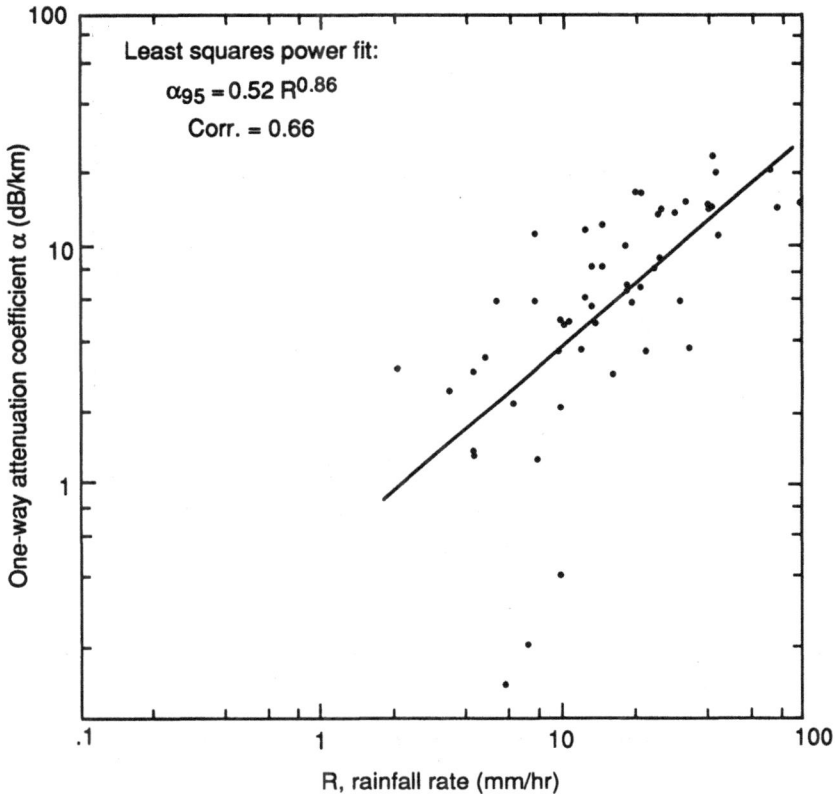

Figure 3.23 Measured attenuation coefficient α for 95-GHz rain data. (Adapted from Richard and Kammerer [6].)

mid-1970s, which is presented in Table 3.6. The model has appeared to hold up as a best estimate of rain attenuation over the years. Figure 3.30 compares the 95-GHz attenuation model to the TNO data of Figure 3.24 [17]. For higher frequencies, it is suggested that the models based on Richard et al. and Nemarich, Wellman, and Lacombe, given in Table 3.7, be used.

3.2.1.3 Snow

Amplitude Characteristics

Average Values. In recent years, considerable effort has been expended to measure the millimeter wave backscatter from falling snow. As a result, data currently exist

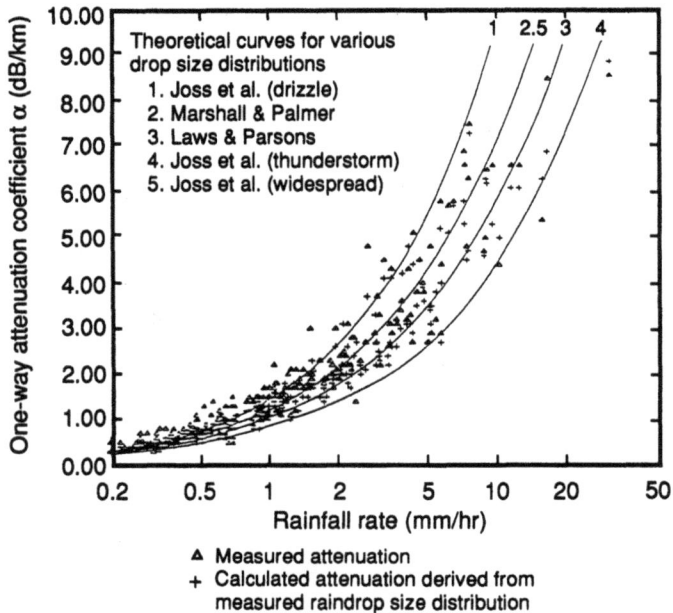

Figure 3.24 Measured and calculated attenuation coefficient α for 94-GHz rain data. (Adapted from Keitzer, Sneider, and de Hann [17].)

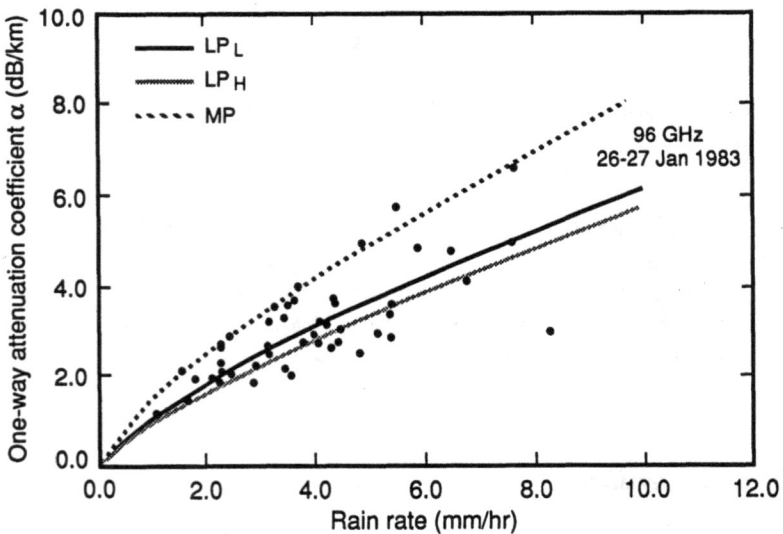

Figure 3.25 Measured and calculated attenuation coefficient α for 96-GHz rain data. (Adapted from Nemarich, Wellman, and Lacombe [18].)

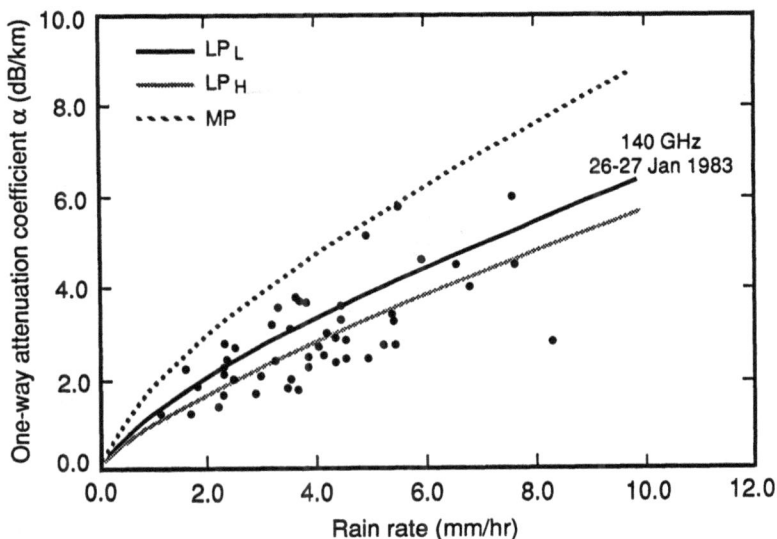

Figure 3.26 Measured and calculated attenuation coefficient α for 140-GHz rain data. (Adapted from Nemarich, Wellman, and Lacombe [18].)

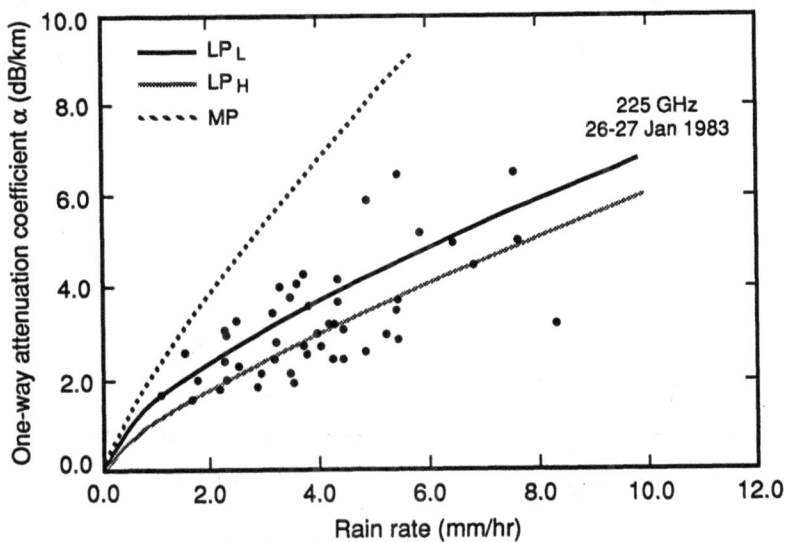

Figure 3.27 Measured and calculated attenuation coefficient α for 225-GHz rain data. (Adapted from Nemarich, Wellman, and Lacombe [18].)

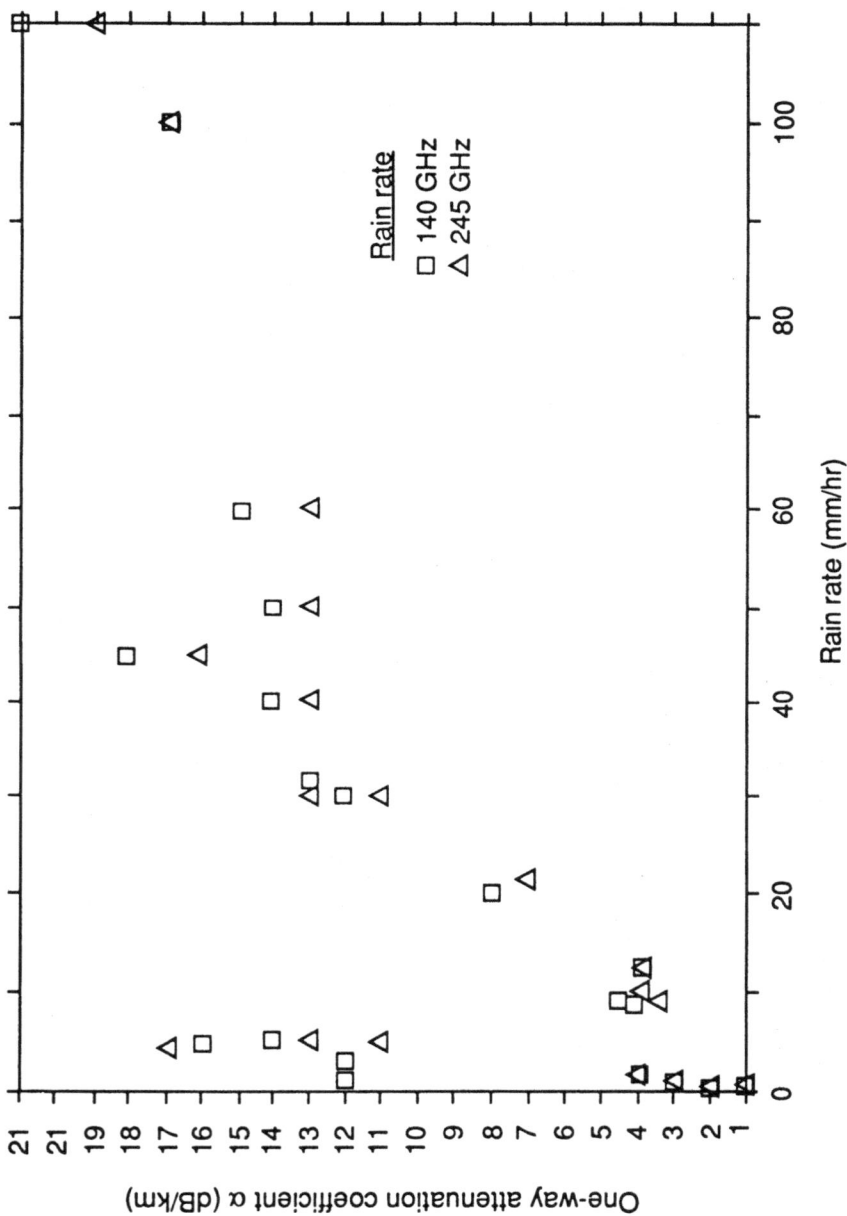

Figure 3.28 Measured attenuation coefficient α for 60- and 81-GHz rain data. (Adapted from V. Furuhama et al. [19].)

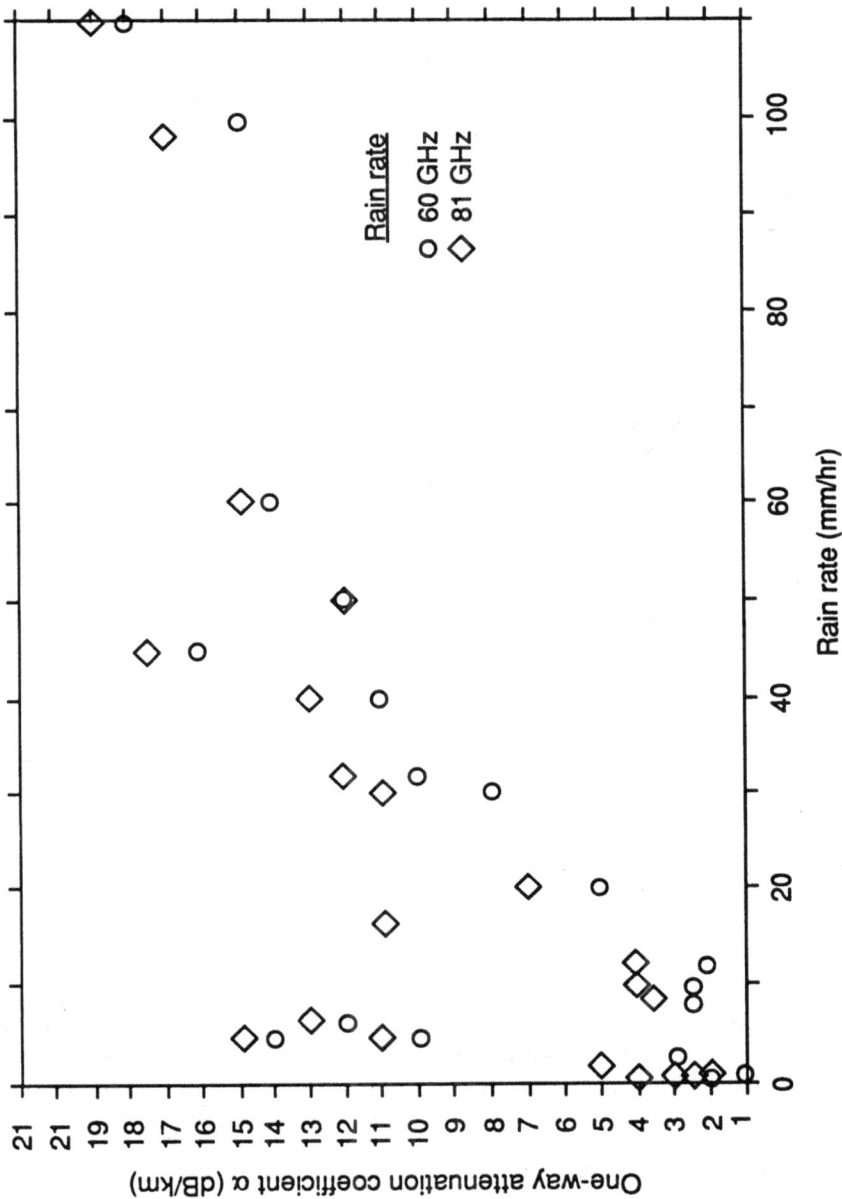

Figure 3.29 Measured attenuation coefficient α for 140- and 225-GHz rain data. (Adapted from V. Furuhama et al. [19].)

Table 3.6

Georgia Tech Attenuation Model for Rain (From Currie, Dyer, and Hayes [5])

$f =$	$10\ GHz^{[1]}$	$35\ GHz^{[2]}$	$70\ GHz^{[3]}$	$95\ GHz^{[4]}$
$\lambda =$	3.2 cm	0.86 cm	0.43 cm	0.32 cm
$\alpha =$	$0.00919R^{1.16}$	$0.273R^{0.985}$	$0.634R^{0.868}$	$1.6R^{0.64}$
$R(mm/hr)$		$\alpha(dB/Km)$		
5	0.0509	1.33	2.56	4.48
10	0.133	2.64	4.68	6.984
15	0.213	3.93	6.65	9.05
20	0.297	5.22	8.54	10.88
25	0.385	6.50	10.36	12.55
30	0.475	7.78	12.14	14.11
35	0.568	9.06	13.88	15.57
40	0.663	10.33	15.58	16.96
45	0.760	11.60	17.26	18.29
50	0.859	12.87	18.91	19.56
55	0.960	14.14	20.55	20.79
60	0.062	15.40	22.16	21.99
65	0.165	16.67	23.75	23.14
70	1.270	17.93	25.32	24.27
75	1.375	19.19	26.89	25.36
80	1.482	20.45	28.44	26.43
85	1.590	21.71	29.98	27.48
90	1.700	22.97	31.50	28.50
95	1.809	24.22	33.02	29.50
100	1.920	25.48	34.52	30.49

[1]Lin and Ishimaru, University of Washington, 1971.
[2]Mueller-Sims, Illinois Water Survey, 1969 Florida Rain.
[3]Lin and Ishimaru, University of Washington, 1971.
[4]J. de Bettencort, Raytheon Company, 1973.

from 35 GHz up through 220 GHz, although the 35-GHz data are not as prevalent as the higher frequency data. Nemarich et al. [22] have been leaders in the collection of this type of data. Figures 3.31 through 3.33 present the radar reflectivity of falling snow at 95, 140, and 225 GHz as a function of the snow concentration (the average weight of snow within a given cubic meter of air at a given time, in gm/m^3). Hand-fit curves are plotted against the raw data. It is desirable to be able to compare the snow data to rain data. In order to accomplish this, we must relate the snow concentration to equivalent rain rate.

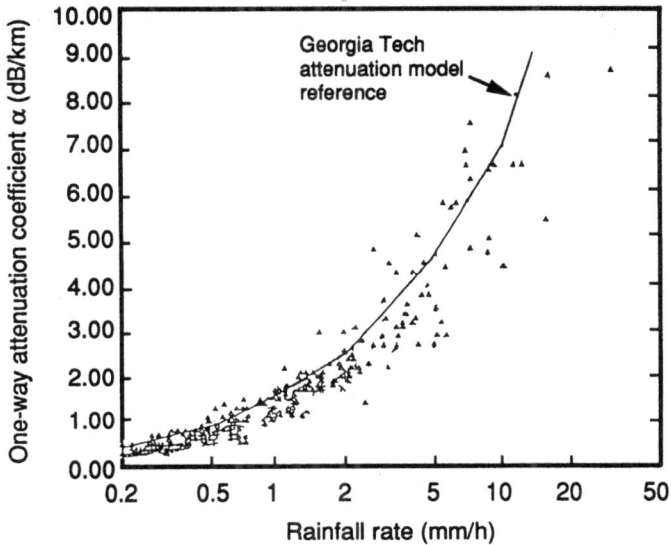

Figure 3.30 Comparison of the Georgia Tech rain attenuation factor model to measured attenuation data for 95 GHz. (From Trebits, © 1989 by Artech House, Inc.)

The equivalent rain rate can be computed from the snow concentration if the snowfall rate is known. Using data from Langelben [23] as in Table 3.8, the relationships between snow concentration and rain rate can be calculated [24].

Nemarich et al. [22] affirmed a conversion factor of 3.3 when comparing rain and snow data. Using a conversion factor of 3.3 mm/hr equivalent rain rate per 1 gm/m^3 of snow concentration, the hand-fit curves to the snow data given in Figures 3.31 through 3.33 are plotted versus rain rate and compared to the LSF curve for the Georgia Tech 95 GHz rain backscatter data shown in Figure 3.34. As can be

Table 3.7
Suggested Attenuation Models for Above 95 GHz

Frequency (GHz)	$\alpha = AR^B$		Source
	A	*B*	
140	1.2	0.75	Richard et al. [21]
225	1.5	0.67	Nemarich et al. [18]

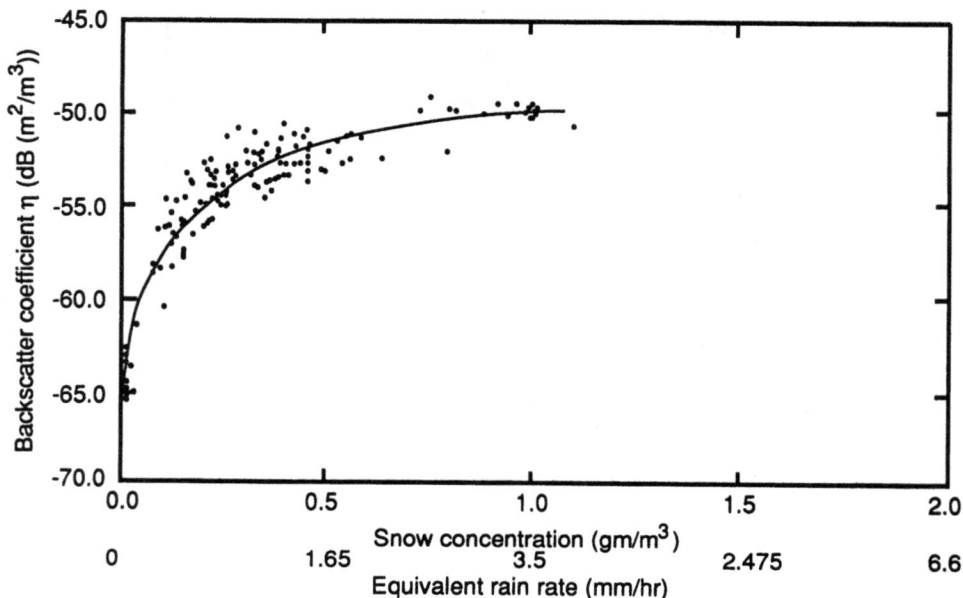

(5) January 1984

Figure 3.31 Airborne snow backscatter data at 95 GHz and hand-fitted curve. (Adapted from Nemarich et al. [22].)

seen, all of the snow data at 95 GHz and higher are lower in reflectivity than the rain at 95 GHz, when plotted versus equivalent rain rate.

Models. Currently, no snow backscatter models are known to exist. However, Nemarich et al. [22] give a summary of data for a snow concentration of 0.5 gm/m³ (roughly 1.6 mm/hr equivalent rain rate), which is reproduced in Table 3.9.

Table 3.10 gives the constants for the equations plotted in Figure 3.33.

3.2.1.4 Sand, Dust, and Debris

Amplitude Characteristics

Average Values. There has been significant work performed on the backscatter and attenuation of sand, battlefield dust, and debris at millimeter wavelengths in recent years, but, unfortunately, most of the reported results are either classified or are

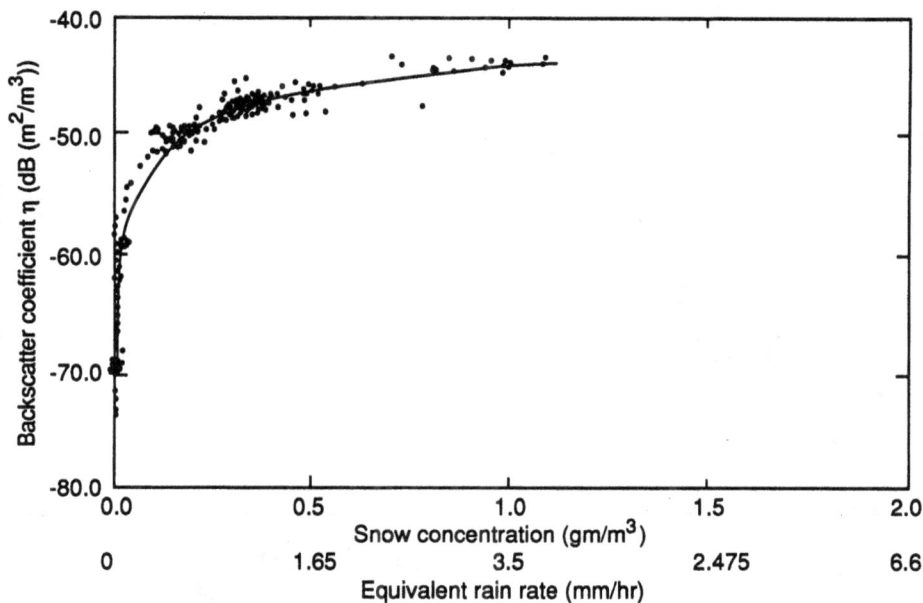

(B) 5 January 1984

Figure 3.32 Airborne snow backscatter data at 140 GHz and hand-fitted curve. (Adapted from Nemirich et al. [22].)

restricted in distribution. Of particular interest are the Smoke/Obscurants Symposia that were conducted throughout the 1980s. Unfortunately, the data obtained in these reports cannot be presented here. However, several papers and reports were presented at open meetings or were approved for public release. The results of these efforts will be summarized here.

One of the landmark experiments concerning millimeter wave reflectivity and attenuation of dust was performed under the Misers Bluff program, in which 800 tons of ammonium nitrate were ignited to simulate a large explosion of magnitude similar to that of a small nuclear burst. Several papers were presented on the reflectivity of the ejected soil and sand. Figure 3.35 gives the backscatter coefficients at 9 and 35 GHz as a function of time in seconds after detonation. As can be seen, a large value for η (approximately 10^{-3}) was recorded at 35 GHz, 2 sec after the blast and exponentially decayed with time. The return at 9 GHz was a factor of 10 lower than that at 35 GHz. Figure 3.36 gives a range profile of the radar return from

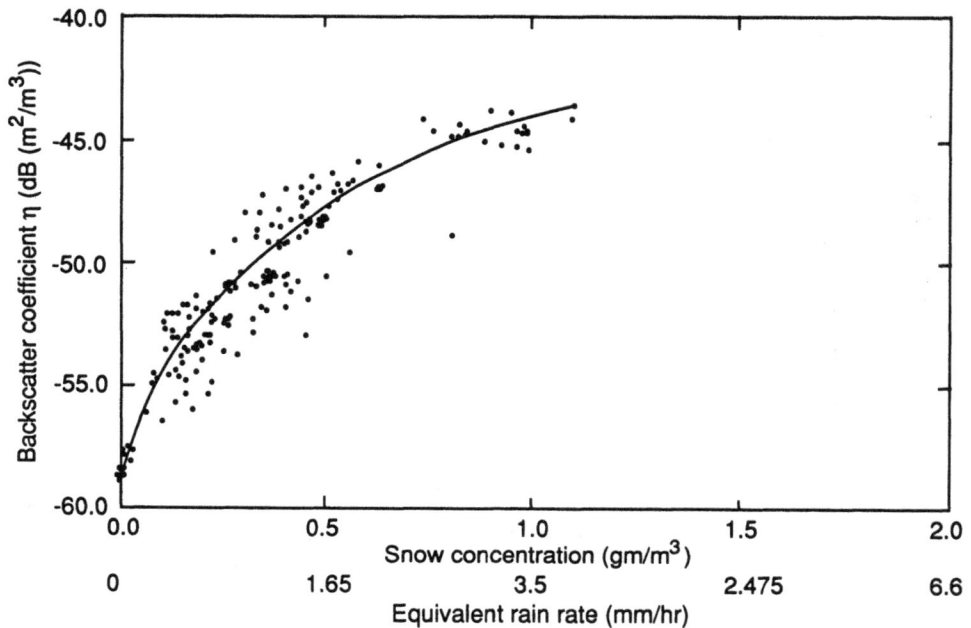

(5) January 1984

Figure 3.33 Airborne snow backscatter data at 225 GHz and hand-fitted curve. (Adapted from Nemirich et al. [22].)

Table 3.8
Relationship Between Snow Concentration and Equivalent Rain Rate [19]

Snow type	Concentration (gm/m³)	Equivalent Rain Rate (mm/hr)
Watery snow	1	3.9
Wet snow	1	3.6
Moist snow	1	3.3
Dry snow	1	3.0

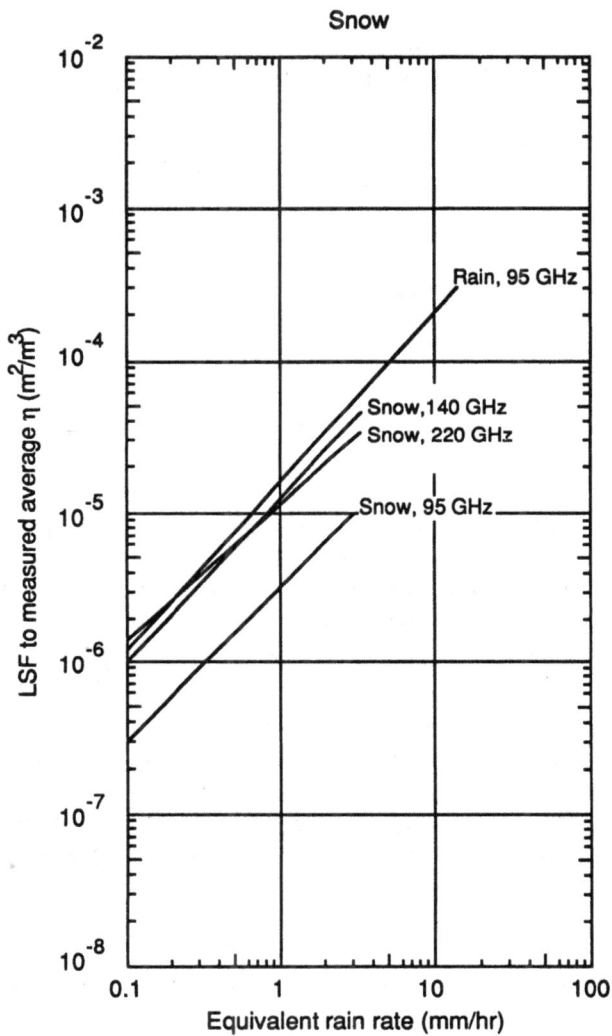

Figure 3.34 Comparison of airborne snow backscatter data at 96, 140, and 225 GHz with Georgia Tech LSF rain backscatter data at 95 GHz.

Table 3.9

Summary of Snow Backscatter per Unit Volume Values (η)
for an Airborne Snow Concentration of 0.5 gm/m^3

	$\eta(dB(m^2/m^3))$		
Date of Measurement	96 GHz	140 GHz	225 GHz
December 6, 1981	−44	—	−39
December 16, 1981	−47	—	—
January 11, 1982	−50	—	—
December 14, 1983	−47	−42	−41
January 5, 1984	−52	−47	−47

the range collocated with the position of the blast for 9 and 35 GHz at times just before the blast and 2 sec after the blast. The large return at 5 km before the blast was from a mesa bluff. As can be seen, 2 sec after the detonation, the return from the cloud was quite large at 35 GHz, and the return from the bluff is smaller, indicating significant attenuation. Table 3.11 summarizes the peak reflectivity values from the cloud at 9, 35, and 95 GHz as a function time after detonation.

Attenuation. Petito [27] reported values for the transmission loss from dirt lofted by simulated battlefield explosions. The backscatter from detonations of both fired 105- and 155-mm shells and C4 explosives, as well as subsurface explosions, were measured at 95 GHz. The results include maximum values of two-way attenuation of 1.38 dB for the 155-mm, 10.7 dB for the 105-mm, and 13.7 dB for the C4 for

Table 3.10

Coefficients for Snow Backscatter Equations Plotted in Figure 3.33

	$\eta = AR^B(m^2/m^3)$	
Frequency (GHz)	A	B
95	3.1×10^{-6}	1.0
140	1.2×10^{-5}	1.04
225	1.15×10^{-5}	0.9

(a)

(b)

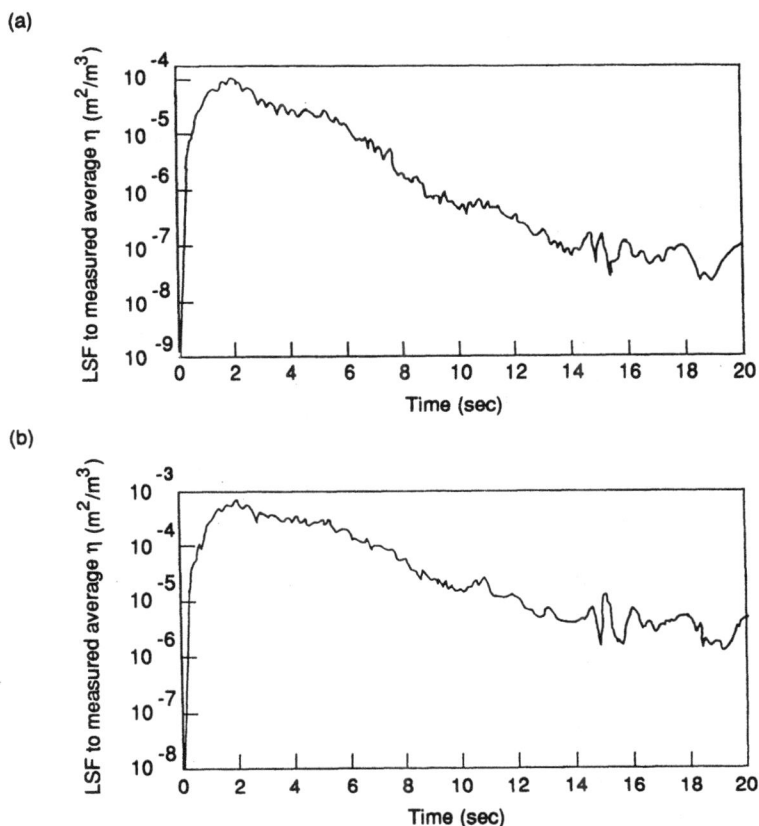

Figure 3.35 Radar reflectivity of dust cloud at Misers Bluff as a function of time after the detonation. (From Martin [25].)

surface detonations. Attenuations of 5.9, 178.8, and 23.5 dB were measured for the 155, 105, and C4 explosives, respectively, for subsurface detonations.

Knox [28] performed measurements of the millimeter wave attenuation of high explosive (HE) dust at Smoke Week II and obtained the data shown in Figure 3.37 for 35, 94, and 140 GHz. The maximum attenuation obtained was less than 0.4 dB. Table 3.12 summarizes the attenuation measurements performed on HE dust. Obviously, the size of the explosions was much smaller than those reported by Petito. Data were also obtained on various smoke, including hexachloroethane, white and

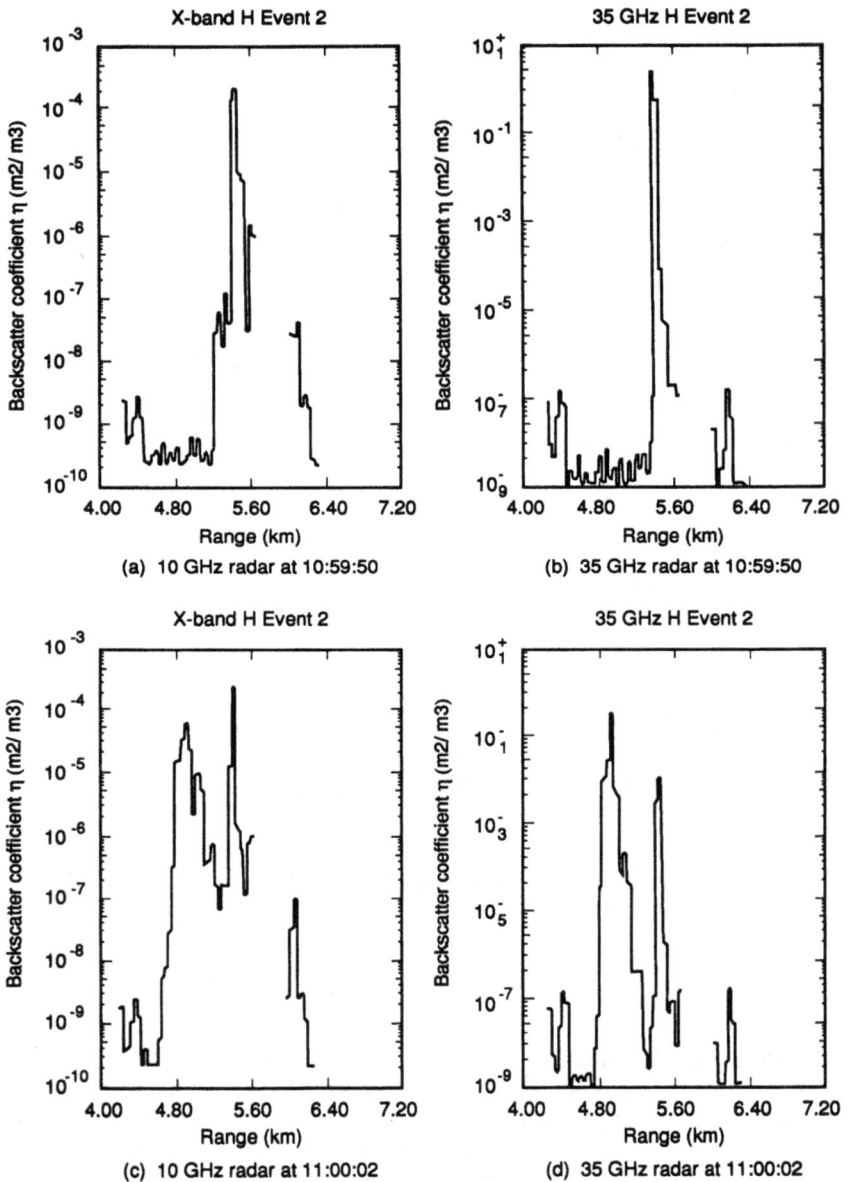

Figure 3.36 Range profile of detonation areas before and 2 sec after the detonation. (From Martin [26].)

Table 3.11

Peak Radar Reflectivity for Dust Cloud at Misers Bluff Event Two (From Martin [26])

Time (sec) after T_o	Antenna Az	Antenna El	Range (meters)	Peak $\eta(m^2/m^3)$ 9.4 GHz	Peak $\eta(m^2/m^3)$ 35 GHz	Peak $\eta(m^2/m^3)$ 95 GHz
−10	73.8	0.4		Pre-event Reference		
2	73.7	0.4	4950	7.3×10^{-5}	2.1×10^{-4}	2.7×10^{-4}
4	73.6	0.4	4950	4.4×10^{-5}	1.2×10^{-4}	1.0×10^{-4}
6	73.8	0.4	4988	2.0×10^{-5}	6.0×10^{-5}	2.2×10^{-4}
8	73.4	0.6	5025	3.3×10^{-6}	3.4×10^{-5}	2.0×10^{-4}
10	73.7	0.8	5023	9.0×10^{-7}	1.2×10^{-5}	—
12	73.7	0.9	5025	8.9×10^{-7}	6.7×10^{-6}	—
14	73.6	1.3	5025	1.5×10^{-7}	3.1×10^{-6}	—
16	73.0	1.4	5025	2.9×10^{-7}	6.7×10^{-6}	—
18	73.3	1.3	5025	1.0×10^{-7}	3.8×10^{-6}	—
20	73.1	1.5	5025	1.5×10^{-7}	3.6×10^{-6}	—
40	72.2	2.8	4988	5.9×10^{-8}	2.5×10^{-6}	—
42	71.6	1.8	4950	2.3×10^{-7}	1.1×10^{-4}	—
44	72.5	2.8	5063	1.9×10^{-7}	2.8×10^{-6}	—
94	73.4	11.2	5325	$4.3 \times 10^{-9}(?)$	1.9×10^{-7}	—

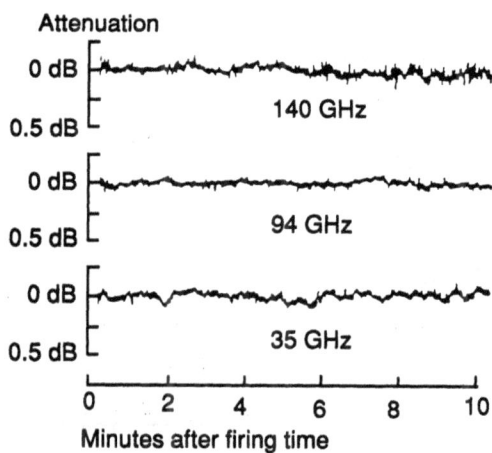

Figure 3.37 Comparison of attenuation due to high explosive dust at 96, 140, and 225 GHz. (From Knox, © 1979 by IEEE.)

Table 3.12
Measured Values of Attenuation for Airborne High Explosive Dust
(From Knox, © 1979 by IEEE [28])

Trial Number	Amount of HE			Maximum Attenuation 35 GHz (dB)	94 GHz (dB)	140 GHz (dB)
23	6 ea.	5 lb	C4	0	0	0
23R	6 ea.	5 lb	C4	0	0.3	0.2
26	6 ea.	15 lb	C4	0.3	0.2	0.4
29	6 ea.	15 lb	C4	0	0	0
30	6 ea.	5 lb	C4	0.3	0	0.3

red phosphorus, oil fog, plasticized phosphorus, white phosphorus wedge, and red phosphorus wedge. Typical results for the red phosphorus smoke attenuation are as shown in Figure 3.38. Again, losses of less than 0.5 dB were observed. Knox reported that "there were no obscurants at Smoke Week II that would have prevented a millimeter wave sensor from operating."

Models. The authors are not aware of any published models for airborne sand attenuation in the open literature. However, Hayes [29] has performed calculations for

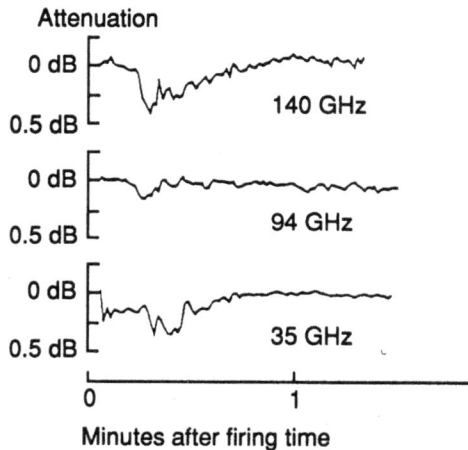

Figure 3.38 Attenuation due to red phosphorus smoke (Wedge). (From Knox, © 1979 by IEEE.)

Table 3.13

Calculated Values of Attenuation Coefficient for Airborne Sand and Soil of Various Types (Adapted from Hayes, ©1989 by RDH, Inc. [29])

Frequency	Material	ε'	$-j\varepsilon''$	$K^2 = \left\|\dfrac{\varepsilon-1}{\varepsilon+2}\right\|^2$	$Im\|-K\|$	$\alpha = \dfrac{8.186\,Im\|-K\|}{\lambda\rho}$ dB/km
3 GHz	fuzed	3.80	0.00038	0.23306	3.3888×10^{-5}	1.06×10^{-5}
10	quartz	3.80	0.00038	0.23306	3.3888×10^{-5}	3.56×10^{-5}
25		3.78	0.000945	0.23133	8.4859×10^{-5}	2.23×10^{-4}
35		4.5	0.0002	0.28994	1.4201×10^{-5}	5.22×10^{-5}
10	fuzed	3.78	0.0006425	0.23133	5.7695×10^{-5}	6.06×10^{-5}
35	silica	3.8	0.0002	0.23306	1.7836×10^{-5}	6.55×10^{-5}
1	calich	2.8	0.0672	0.14079	8.7483×10^{-3}	2.80×10^{-3}
10	clay	2.4	0.0576	0.10139	8.9241×10^{-3}	9.51×10^{-3}
3	dry	2.55	0.01681	0.11606	229.1×10^{-5}	72.21×10^{-5}
10	sandy soil	2.53	0.00911	0.11410	133.18×10^{-5}	1.4×10^{-3}
3	3.88% x	4.40	0.2024	0.28294	1.4809×10^{-2}	4.66×10^{-3}
10	sandy soil	3.60	0.4320	0.22020	4.1082×10^{-2}	4.32×10^{-2}
1	2.74% w	6.60	0.3366	0.42489	1.3632×10^{-2}	1.49×10^{-3}
10	sandy soil	4.80	0.2064	0.31292	1.3379×10^{-2}	1.46×10^{-2}
3	water	76/77	12.04	0.92693	5.6984×10^{-3}	4.66×10^{-3}
10	25°C	55	29.70	0.91939	21.5682×10^{-3}	5.88×10^{-2}
25		34	9.01	0.84969	19.2706×10^{-3}	13.39×10^{-2}

the attenuation of airborne sand and clay at millimeter wavelengths. His equation is of the form

$$\alpha = \frac{8}{3} \pi k^4 a^6 |K|^2 N' \qquad (3.9)$$

where $k = 2\pi/\lambda$ and N' is the expected concentration of particles at the radius which describes the average mass concentration of sand. Table 3.13 summarizes some attenuation values at 3 to 35 GHz for various materials.

3.3 SURFACE CLUTTER

3.3.1 Land Clutter Characteristics

3.3.1.1 Average Values

Overview. Significant efforts have been dedicated to measuring and modeling the radar backscatter of natural surface clutter at millimeter waves, but most of the efforts have concentrated at 35 and 95 GHz, the two frequencies of primary interest at present for millimeter radar applications. Georgia Tech has developed a model for millimeter land backscatter over the last 20 years, with periodic updates as new data became available [26, 27, 28, 29]. Figure 3.39 summarizes the millimeter-wave data available in the mid-1970s as a function of wavelength. Below 2 cm (above 15 GHz), a relatively strong frequency dependence in the reflectivity is observed in the data. More recent data confirm a strong frequency dependence in the data. In the following sections, summaries of measured data are presented along with the predicted curve from the Georgia Tech land clutter model presented in [29].

Grass, Crops, and Trees. Several measurement programs over the last 15 years have collected data on the surface reflectivity of grass crops and trees. Figure 3.40 summarizes the data from several types of vegetation over the grazing angles of 0° to 70° [31, 32]. The data from [31] were collected from an airborne platform in Western Europe. As can be seen, at millimeter wavelengths, the differences in the maximum and minimum returns from various types of vegetation vary by only a few decibels. The plateau region obviously covers the plotted angles of 12° and 70°, and thus little variation in $\sigma°$ with grazing angle is observed. (See Sec. 2.4.1 for a discussion of the angular dependence of rough surfaces.)

Figures 3.41 through 3.43 present data on the reflectivity of grass and crops and trees at 35 and 95 GHz compared with the Georgia Tech land clutter model. For the grass and crops data shown in Figures 3.41 and 3.42, the critical angle (angle

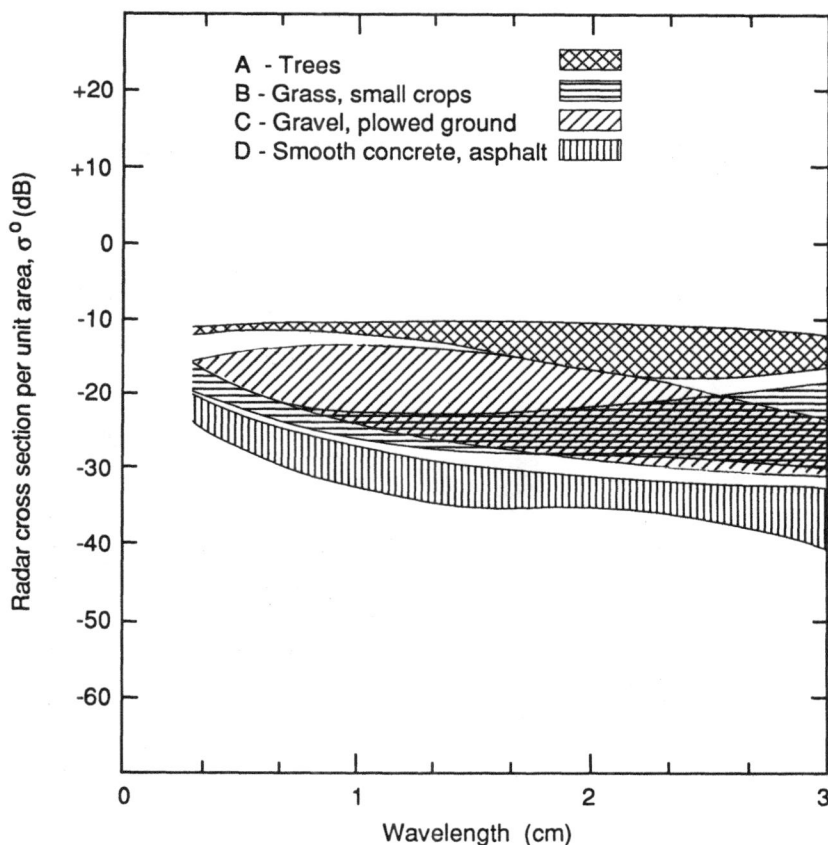

Figure 3.39 Clutter radar reflectivity values (σ^0) between 20° and 70° grazing angle as a function of wavelength. (From Dyer and Hayes [30].)

for which the rough surface starts to look smooth) appears to be about 15° at 35 GHz and 10° at 95 GHz. Otherwise, there is little difference in the reflectivity between the two frequencies. Figures 3.43 and 3.44 show that the critical angle for trees at 35 and 95 GHz is very low, as would be expected since the surface roughness for trees is very high compared to grass. Greater angular dependence is exhibited by the 95-GHz data than the 35-GHz data, which would not be predicted by theory. Chances are that the high points are caused by manmade or natural "specular" targets that may have been present in the data unbeknownst to the experimenter. These targets would appear much larger at 95 GHz than at 35 GHz and illustrate one of

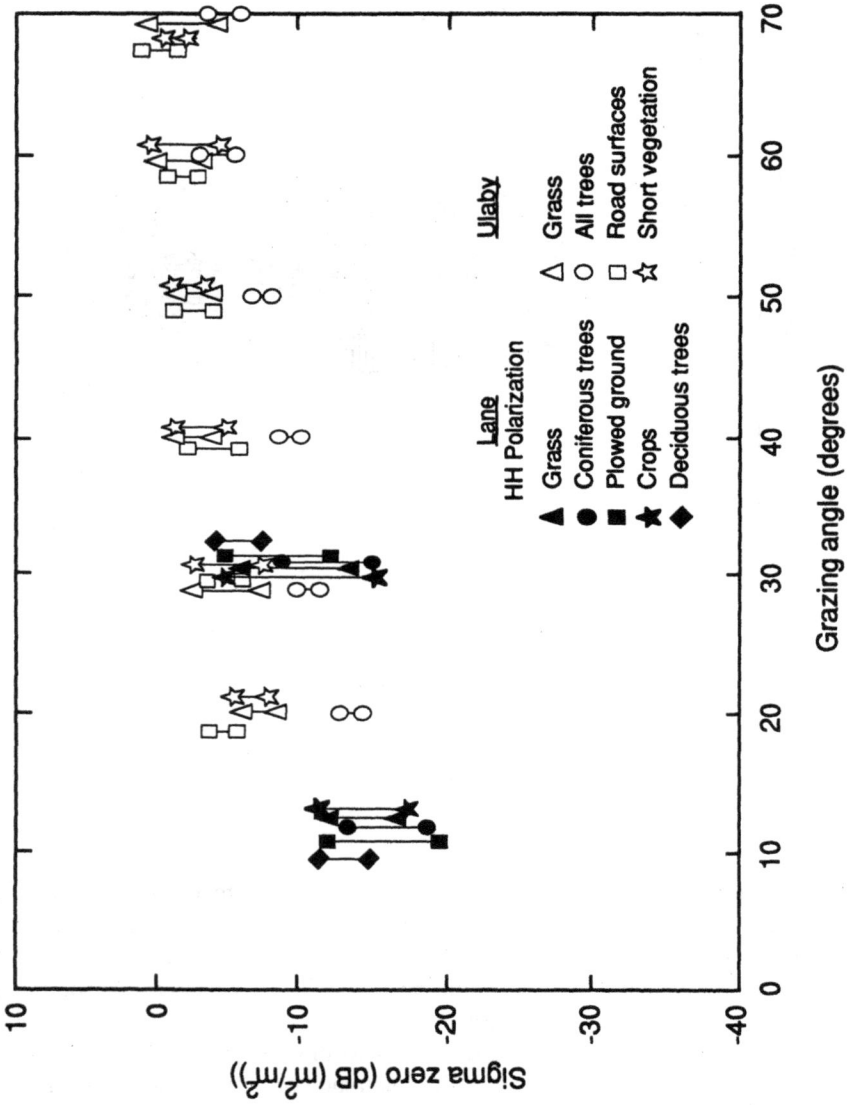

Figure 3.40 Surface vegetation radar reflectivity (σ^0) values between 10° and 70° grazing angle.

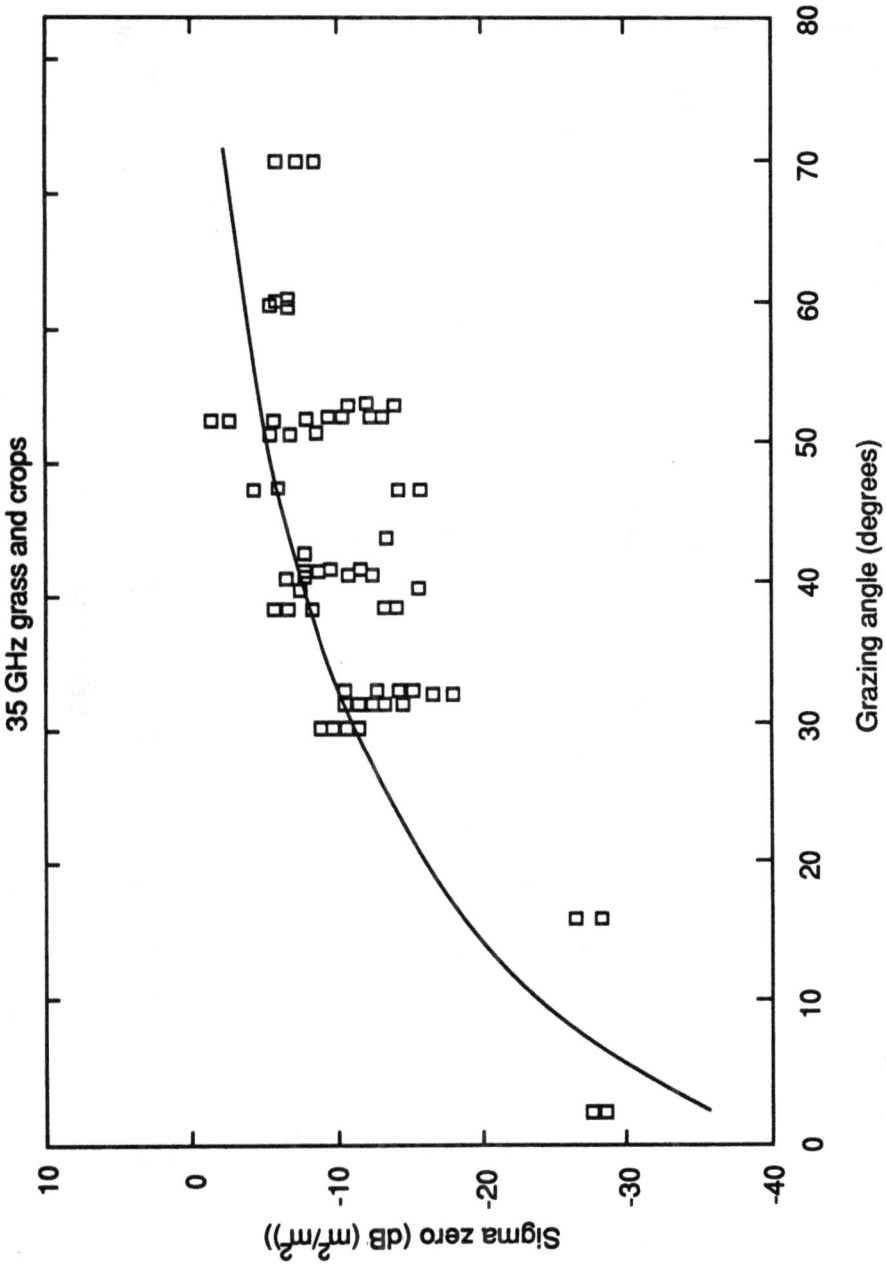

Figure 3.41 Reflectivity of grass and crops between 0° and 70° grazing angle for 35 GHz.

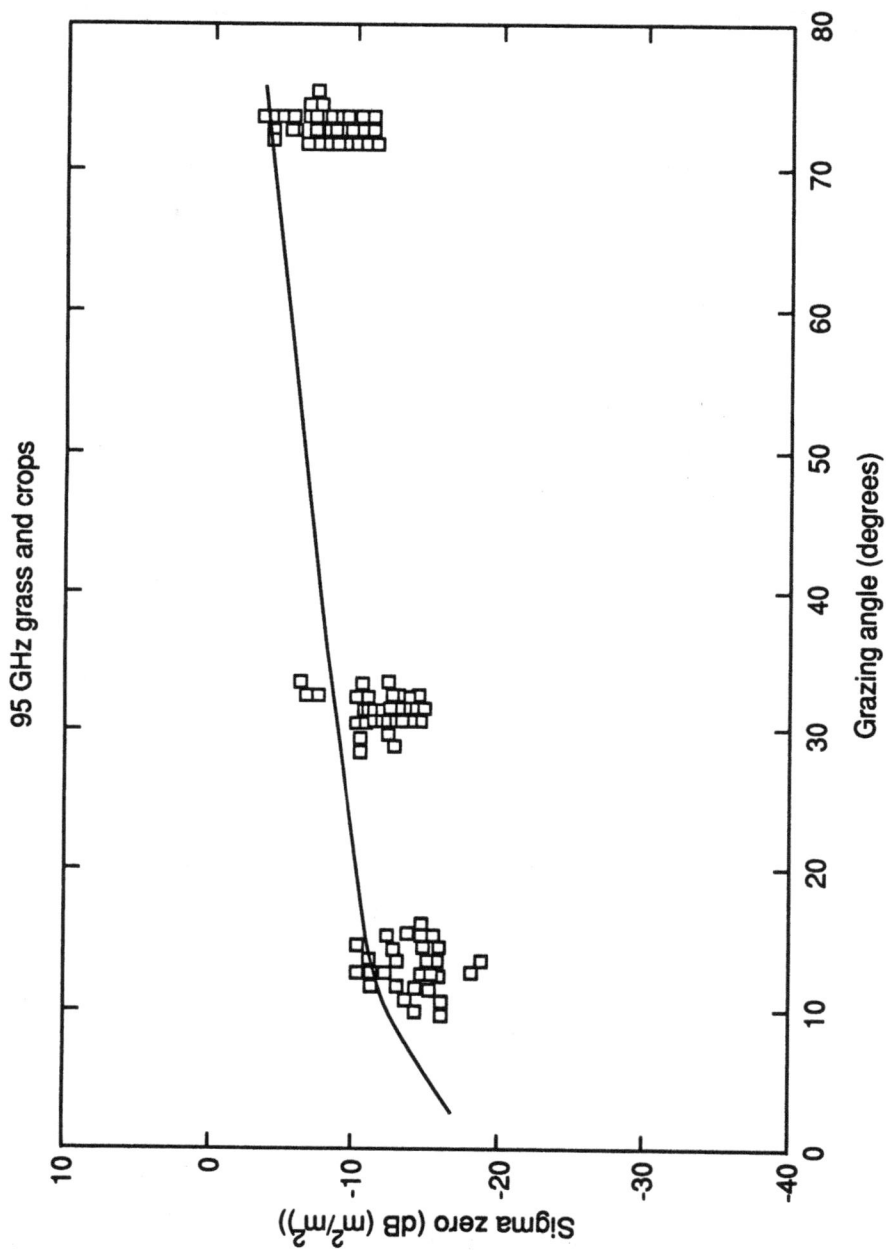

Figure 3.42 Reflectivity of grass and crops between 0° and 70° grazing angle for 95 GHz.

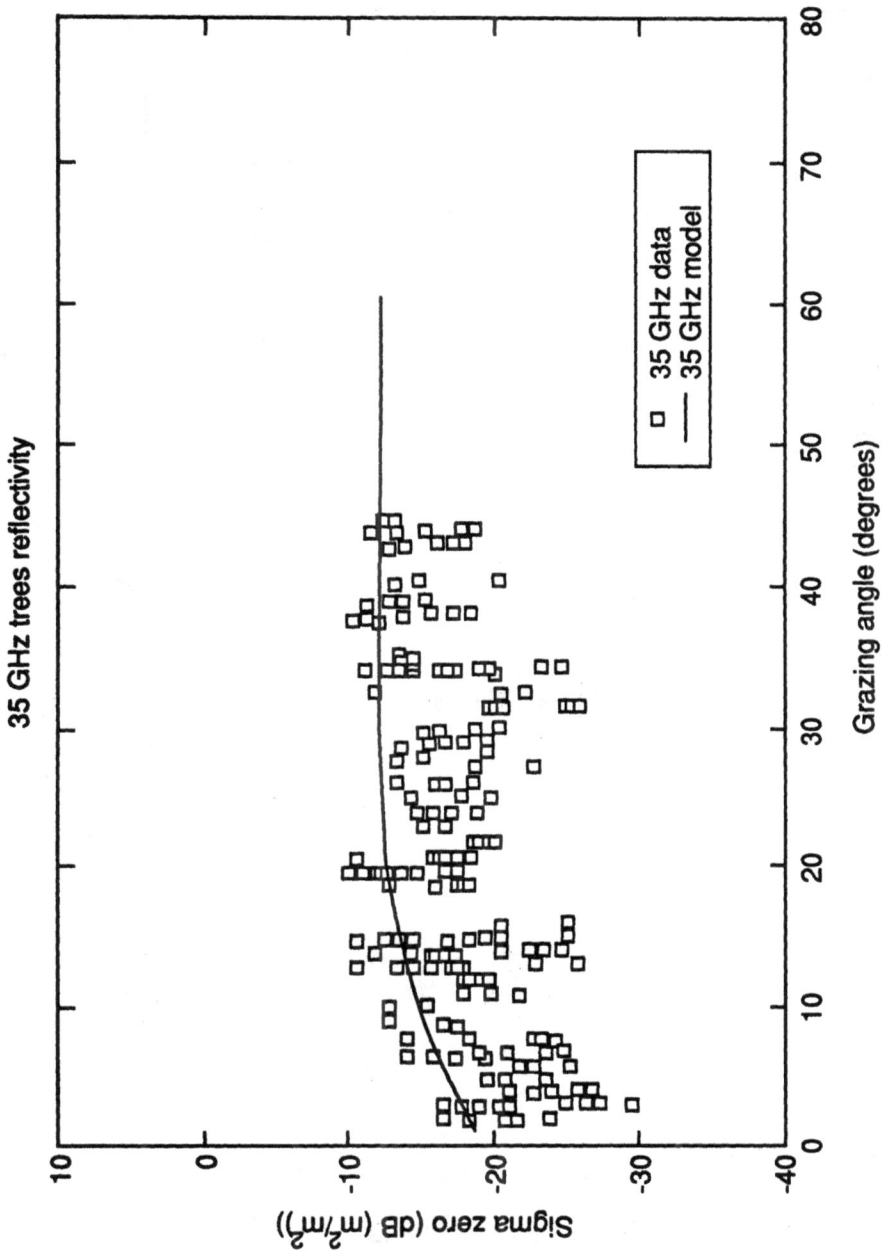

Figure 3.43 Reflectivity of trees between 0° and 70° grazing angle for 35 GHz.

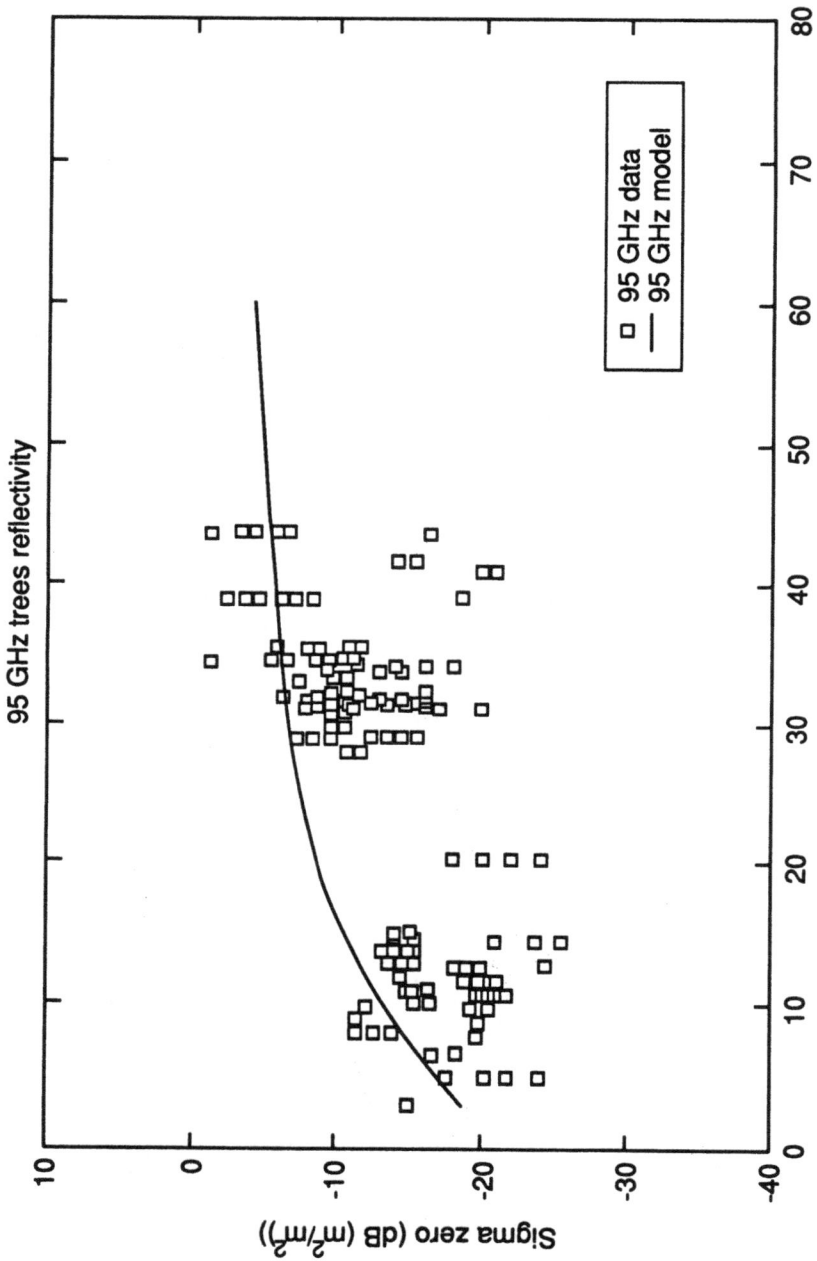

Figure 3.44 Reflectivity of trees between 0° and 70° grazing angle for 95 GHz.

the major problems in detecting targets in clutter at millimeter waves: the point targets immersed in the clutter that almost always cause excessive false alarms for standard constant false alarm rate (CFAR) processing (see Sec. 4.2.1.3).

Snow-Covered Ground. The reflectivity of snow-covered terrain has been a puzzle for many years. Airborne measurements of snow-covered terrain have upon occasion produced higher than expected values, which caused some experimenters to question their equipment (or in some cases their sanity). In 1976, the U.S. Air Force sponsored a program to determine the reason for such erratic measurements of snow reflectivity at 35 GHz [33]. The results indicated that the reflectivity of snow changed significantly whenever the snow went through a melt/freeze cycle, being low when wet and high after refreezing. Also, it was determined that the amount of liquid water in the snow was a strong indicator of backscatter intensities being low when the liquid water content was high and high when the liquid water content was low.

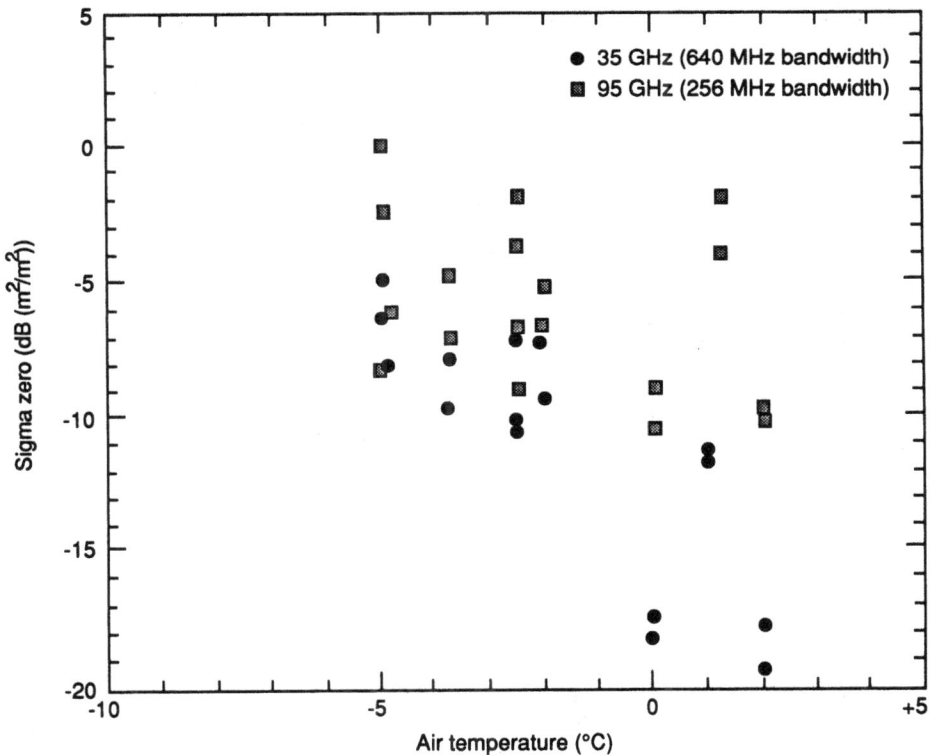

Figure 3.45 Reflectivity of snow-covered ground for 35 and 95 GHz as a function of air temperature, (10° to 30° grazing angle). (From Currie et al. [36].)

146

These results were confirmed in 1977 when measurements were repeated at the same site for frequencies of 8, 18, and 35 GHz [34]. These measurements indicated that the same effect was observable at lower frequencies as well as at 35 GHz. As a result of these and other experiments, data from snow-covered ground tend to be separated into two groups: wet and dry or wet and refrozen, with correspondingly different reflectivity values. Later experiments illustrated that the same trend occurred at 95 GHz as well [35]. Figures 3.45 and 3.46 give the snow reflectivity at 35 and 95 GHz averaged over a wide bandwidth to remove spatial variations as a function of air temperature and liquid water content, respectively [36]. As can be seen from the data in the figures, the snow reflectivity rapidly decreases as the air temperature approaches 0°C and as the liquid water content approaches 7% by weight. Both measurements, air temperature above 0°C and high liquid water content, indicate significant melting in the snow.

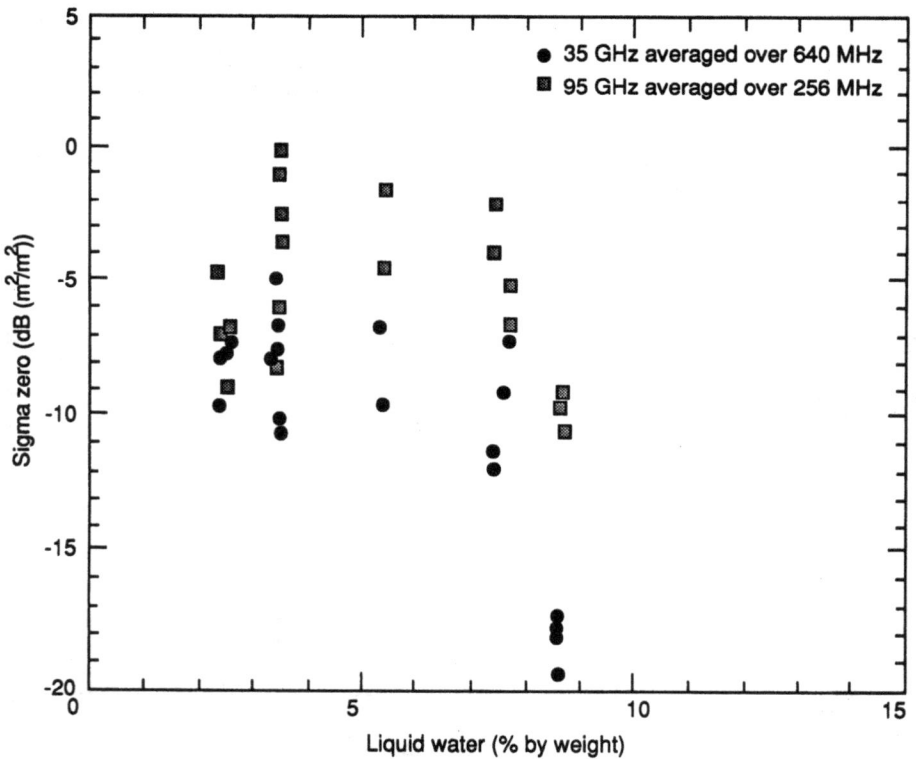

Figure 3.46 Reflectivity of snow-covered ground for 35 and 95 GHz as a function of liquid water content (10° to 30° grazing angle). (From Currie et al. [36].)

Figures 3.47 and 3.48 give the reflectivity of snow-covered ground for wet and refrozen (dry) snow conditions at 35 and 95 GHz compared with the prediction of the Georgia Tech model. The wet snow data are 8 to 10 dB lower than the refrozen snow data at 35 GHz, and the 95-GHz wet snow data are 4 to 5 dB higher than the 35-GHz snow data. The 95-GHz refrozen snow data are 10 dB higher than the 95-GHz wet data. The 95-GHz refrozen snow data exceed 0 dB values for σ^0 above 20° grazing, and are likely to represent a severe detection environment for many types of military targets.

Figure 3.49 compares the reflectivity of snow-covered buildings and snow-covered trees to snow-covered ground for circular polarization (RL and LL). The data were collected from an airborne sensor which tracked the ground. It is interesting that the reflectivity of the trees was less than that of the snow-covered ground, the trees appearing more like tree data without snow cover (given in Figure 3.44). Although specific liquid water content data were not available for these measurements, the time of day and air temperature would indicate that the data in Figure 3.49 are wet snow as opposed to dry (refrozen) snow.

Model for Surface Scatter. An empirical model has evolved at Georgia Tech for millimeter-wave land clutter reflectivity over many years. The most recent version of the model assumes an exponential dependence of σ^0 on depression angle and takes the form [37, 38, 39]:

$$\sigma^0 = A(\theta + C)^B \qquad (3.10)$$

where

A, B, and C	are empirically derived constants that depend on frequency and clutter type
σ^0	is the RCS per unit area in dB (m^2/m^2) (backscatter coefficient)
θ	is the depression angle in degrees ($5° < \theta < 70°$)

Table 3.14 gives the parameters for several clutter types at 35 and 95 GHz, including crops and grass, trees, and snow-covered ground. Note that two separate models are presented for snow, wet and refrozen. This is because refrozen snow is significantly higher in backscatter than wet snow, as discussed previously. Although the data in Figures 3.41 through 3.49 were used to develop the model, it appears to correlate well with independent data sets. As an example, Figure 3.50 compares independent reflectivity data on snow backscatter with the Georgia Tech model. No measurement was made of the snow state, but the data generally fall between the wet and refrozen predictions of the model, which indicates good agreement.

Statistical Reflectivity Model. Starting with the average value for σ^0 from the model in Table 3.14 and assuming that the model in Table 3.14 gives the σ^0 value for a

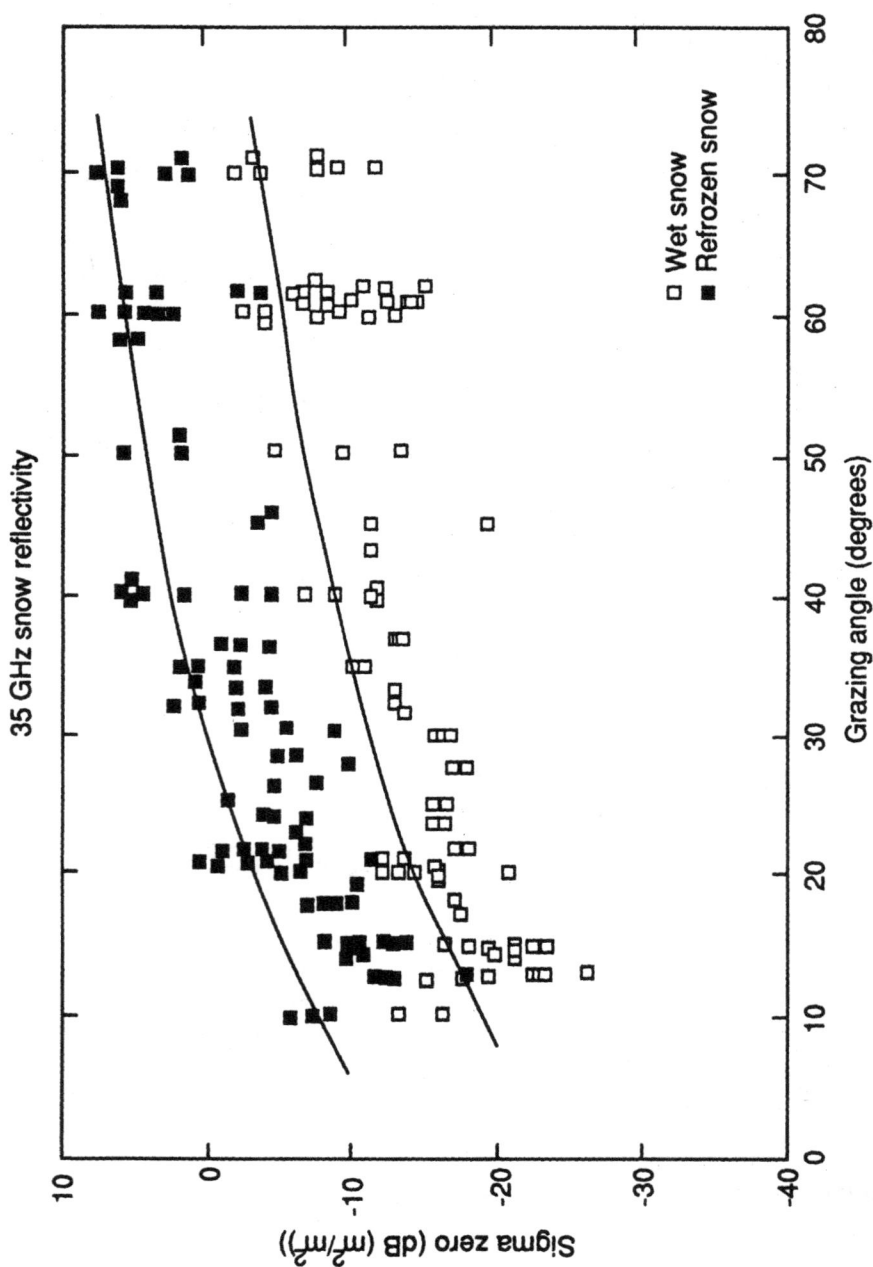

Figure 3.47 Reflectivity of snow-covered ground for 35 GHz compared to the Georgia Tech model prediction, wet and refrozen snow.

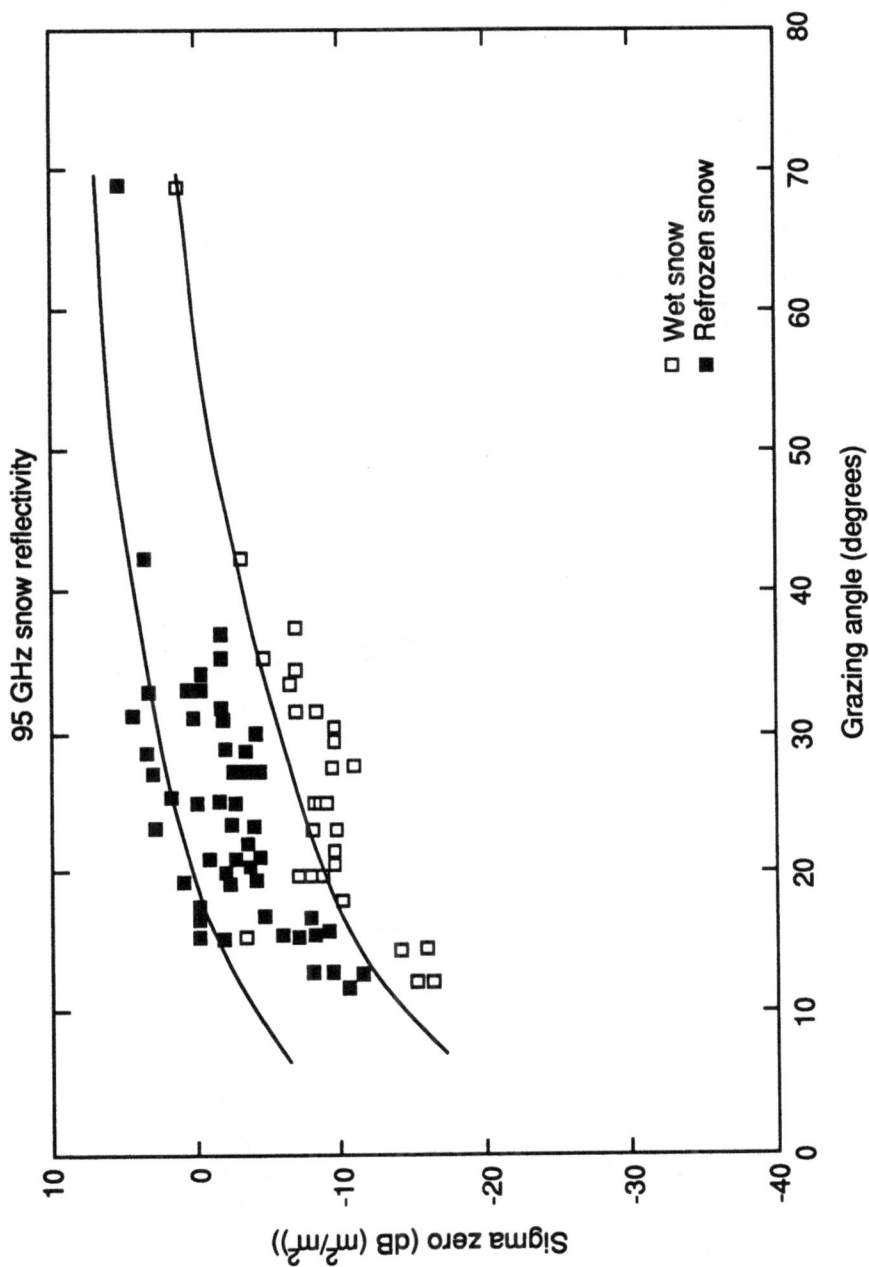

Figure 3.48 Reflectivity of snow-covered ground for 95 GHz compared to the Georgia Tech model prediction, wet and refrozen snow.

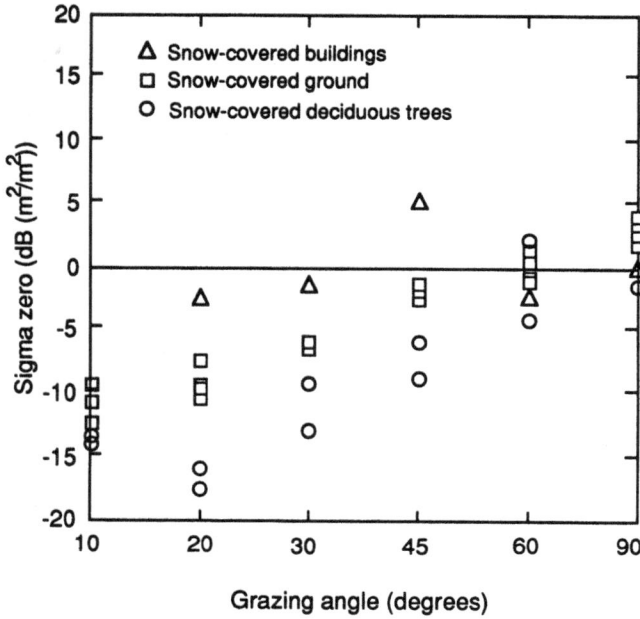

Figure 3.49 Reflectivity of snow-covered buildings, snow-covered trees, and snow covered ground for 95 GHz, circular polarization, wet snow. (From Currie et al., © 1988 by IEEE.)

Table 3.14
Constant Parameters for the Georgia Tech Empirical Millimeter-Wave Clutter Model

Clutter Type	Freq.	A	B	C
Wet Snow	35 GHz	0.23	1.5	0.033
	95 GHz	0.58	1.5	0.037
Refrozen	35 GHz	2.30	1.5	0.035
	95 GHz	3.00	1.5	0.037
Trees	35 GHz	0.047	0.5	0.031
	95 GHz	0.34	1.5	0.035
Grass/Crops	35 GHz	0.30	2.0	0.036
	95 GHz	0.20	1.0	0.036

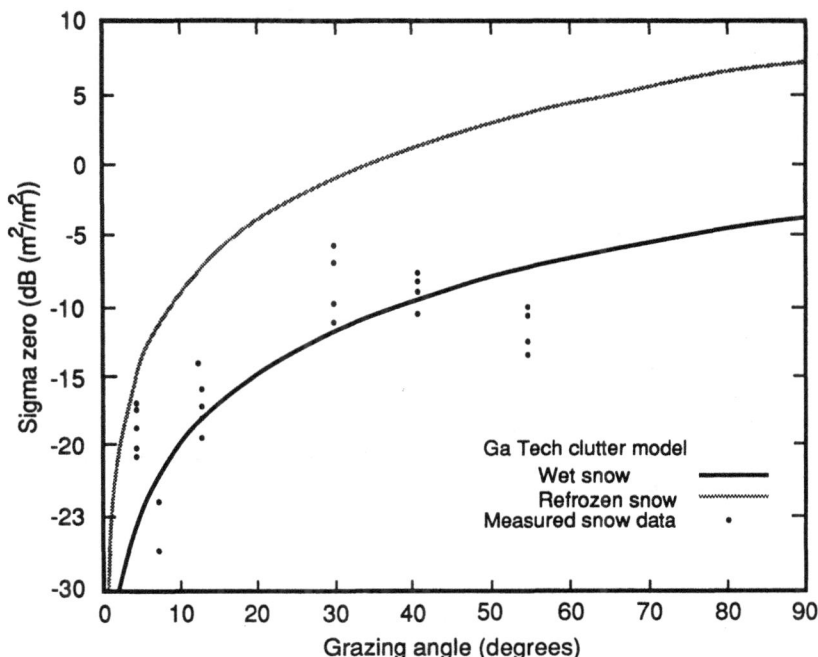

Figure 3.50 Comparison of reflectivity data for snow-covered ground at 35 GHz with the Georgia Tech model predictions for wet and refrozen snow.

specific depression angle, it is possible to reproduce the average of measured data values. However, actual data have a spread in values because of tilt variations from horizontal of the illuminated terrain. If we assume that such slope variations are Gaussian in the down-range plane, with a standard deviation of 0.06 rad and the spatial variation given by a Weibull distribution, then we have

$$\sigma^0 = \overline{\sigma}_m \frac{\Gamma(1 + a)}{(\ln 2)^2} \qquad (3.11)$$

where

σ_m = median value for σ^0
Γ = gamma function
a = $1/b$ (b is the Weibull slope parameter)

A generator of the form $SC \cdot AF \cdot XU^a$ is used to compute the statistics on a microcomputer, where

SC   is the average value
AF   is the reciprocal of $\Gamma(1 + a)$
XU   is the random deviate on the interval $(0,1)$

This method can be used to compute statistics for clutter or as a "Monte Carlo" generator of random clutter samples with the correct statistics. Figures 3.51 and 3.52 give examples of statistics resulting from the model compared with actual data for grass and crops at 35 and 95 GHz. The solid lines represent the tenth, thirtieth, fiftieth, seventieth, and ninetieth percentile statistics generated by the model overlaid on actual data.

Figures 3.53 and 3.54 compare the statistical model predictions with the data for wet and refrozen snow at 95 GHz. Again, the model appears to predict the spread in the data extremely well.

3.3.1.2 Temporal Characteristics

Most ground clutter, such as dirt and snow, does not exhibit temporal variations, although spatial variations are evident. The exception to this rule is vegetation. Since vegetation contains limbs, twigs, and leaves (or needles) that can move with the wind, it can exhibit temporal variations. The following discussion of millimeter-wave temporal characteristics of vegetation is adapted from Hayes and Long [40].

Barlow [41] found from Doppler radar measurements at L-band that the spectra of most types of clutter can be represented by a Gaussian function of the form

$$W(f) = W_0 \exp(-af^2/f_t^2) \tag{3.12}$$

where

W is the power density
W_0 is the power density at zero fluctuations
f_t is the radar center frequency
a is an experimentally determined constant

Barton [42], Skolnik [43], and Long [44] have expanded on previous works and related the constant a with the standard deviation of a Gaussian function and with the measured spectra.

Thus, from

$$W(f) = W_0 \exp(-f^2/2s_c^2) \tag{3.13}$$

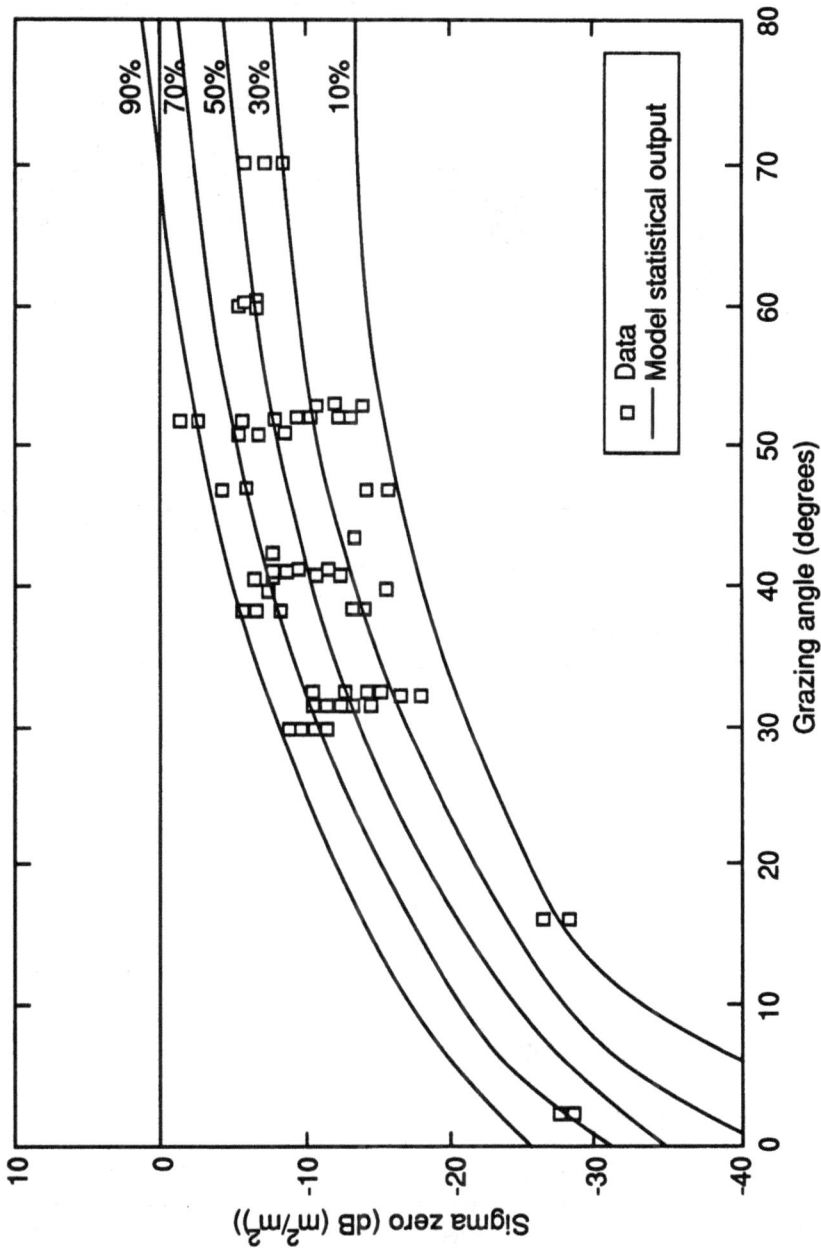

Figure 3.51 Comparison of the reflectivity of grass and crops for 35 GHz with the Georgia Tech statistical model predictions for 10%, 30%, 50%, 70%, and 90% distribution levels.

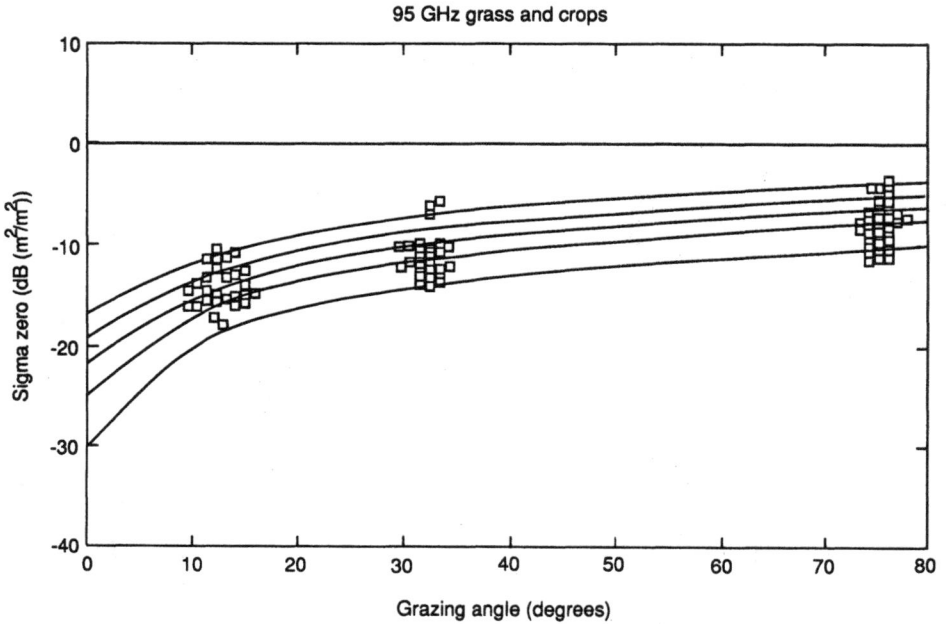

Figure 3.52 Comparison of the reflectivity of grass and crops for 95 GHz with the Georgia Tech statistical model predictions for 10%, 30%, 50%, 70%, and 90% distribution levels.

it can be shown that

$$a = \frac{f_t^2}{2s_c^2} = \frac{f_t^2}{2\left(\dfrac{2s_v}{\lambda}\right)^2} = \frac{c^2}{8s_v^2} \tag{3.14}$$

where

s_c = rms clutter frequency spread
s_v = rms Doppler spread
c = velocity of light

As reported by Long [45], coherent data measured by Billingsley and Larrabee [12] for wind-blown trees at L- and X-bands appear exponential (but not Gaussian) down to the −60-dB level. They reported that the spectrum decays exponentially

95 GHz wet snow reflectivity

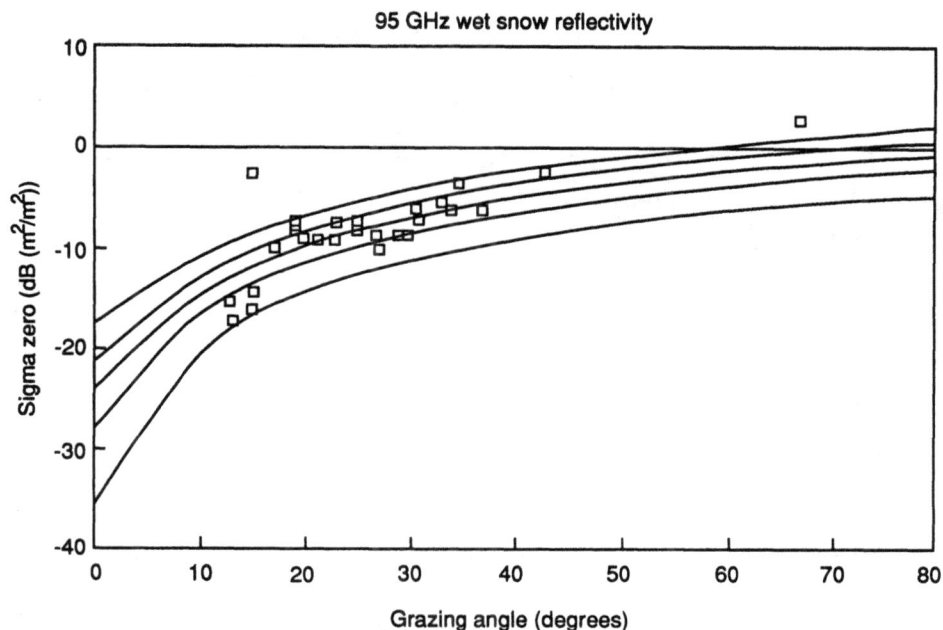

Figure 3.53 Comparison of the reflectivity of wet snow for 95 GHz with the Georgia Tech statistical model predictions for 10%, 30%, 50%, 70%, and 90% distribution levels.

with a linear increase in Doppler velocity and does not exceed 2 m/s at the −60-dB level. However, many experimenters have reported spectral characteristics wider than predicted by an exponential function for *noncoherent* spectra. Nathanson has noted differences in coherent and noncoherent spectra, which include

1. For coherent spectra, the center frequency (peak energy) can be either positive or negative depending on whether the relative motion between radar platform and the scatterers is positive or negative.
2. For noncoherent spectra, the spectra are broadened due to the foldover in spectra because Doppler direction (sign) is not preserved.

In addition, a physical mechanism exists that could account for two or more spectral mechanisms (i.e., a low-frequency exponential shaped spectra due to limb and branch motion and a higher frequency wider than exponential shaped spectra due to leaf flutter). Likewise, for sea clutter, wave motion and spray droplet motion may account for two separate mechanisms. The effects of leaf flutter on the spectra would expect to increase with frequency, particularly at millimeter wavelengths.

95 GHz refrozen snow reflectivity

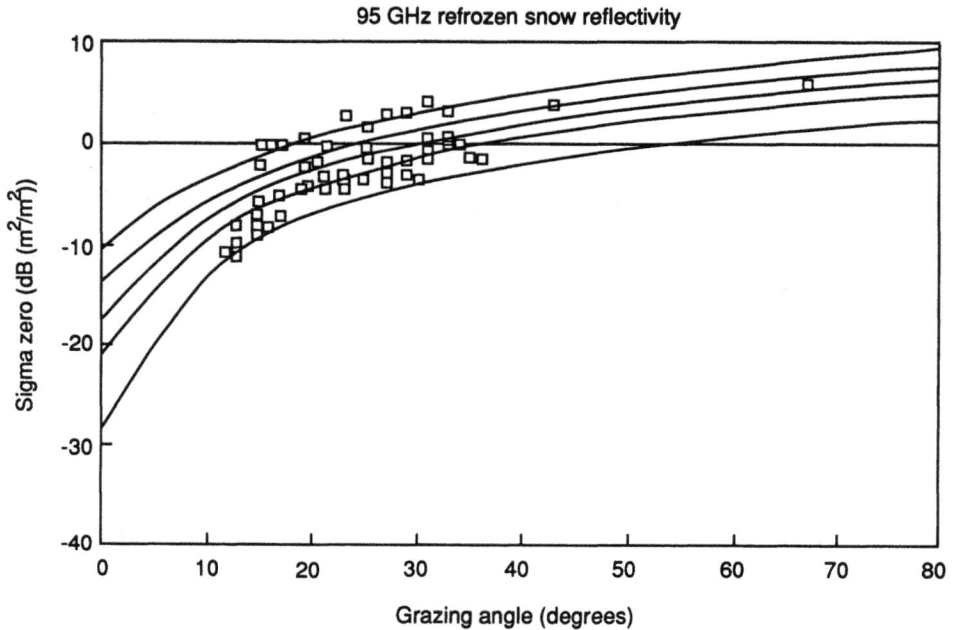

Figure 3.54 Comparison of the reflectivity of refrozen snow for 95 GHz with the Georgia Tech statistical model predictions for 10%, 30%, 50%, 70%, and 90% distribution levels.

Data collected on tree clutter spectra at millimeter wavebands to date have had two problems from the standpoint of determining Doppler spectra shape: they have been noncoherent and have been collected with logarithmic receivers. It is only recently that high-quality linear coherent radar has been available at 35 and 95 GHz, but only target data have been collected with these systems. However, we must work with the data available, and millimeter-wave spectra have been computed on the available noncoherent data. The logarithmic data were "linearized" prior to spectral shape analysis, but there is no question that the shapes may have been affected by the original logarithmic receiver transfer curve. These data indicated the presence of a second nonexponential spectral response for the return from trees that is some 15 to 20 dB below the maximum signal level. The following discussion is based on these data.

As the radar frequency approaches the millimeter waveband, the simple Gaussian power spectral density function (PSDF) may not describe the measured data. When radar signals contain both a slow-varying component and a fast-varying component, as discussed above, then it is important to consider the operation of the signal processor. Previous data measured in the Doppler (or frequency) domain indicate that the low-frequency components around the dc value are quite large compared to the high-frequency components. Thus, a high-pass filter with a low-frequency cutoff will remove a large portion of the total return from the trees. This, of course, has been observed in essentially every MTI radar developed. (They work!)

The fraction of power contained in the slow-varying component can be determined from the equations just presented by considering the fraction of the PSDF area corresponding to the Gaussian function, namely

$$\frac{\text{Slow component of } \sigma^0}{\text{Total } \sigma^0} = \frac{A[2.4853s]}{A[2.4853s + 0.062326f_c]} \tag{3.15}$$

Measured noncoherent data at S-band, C-band, X-band, and K_u-band indicate that the slow-moving component represents 77% of the total magnitude, and the fast-moving component represents 23% of the total magnitude of σ^0. Long [44] considered the polarization matrix generated for X-band tree returns and showed that approximately 70% of the total return was slowly varying and did not depolarize. This is remarkably consistent with the data presented here.

In a like manner, consider the fractional part of the total PSDF represented by the Gaussian portion at K_a-band and M-band; 64% of the signal at K_a-band and 52% of the signal at M-band vary slowly about the dc value. Figure 3.55 shows the form

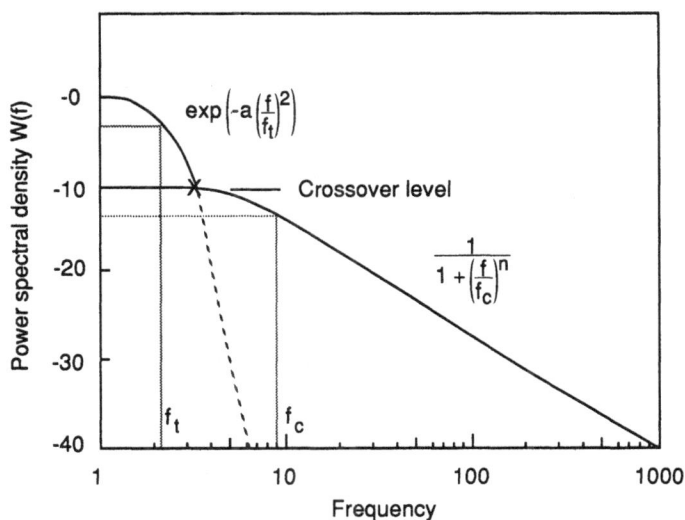

Figure 3.55 Low- and high-frequency components of the noncoherent power spectral density for wind-blown trees.

of the frequency spectra, including the low-frequency Gaussian part and the high-frequency Lorentzian part. Table 3.15 summarizes the measured "break points" for the two types of dependencies.

Considering these divisions of power in the frequency domain, the same ratio can be used in the time domain to generate the complete time correlation functions. Before generating the correlation functions, however, let us first consider the relationship between the time and frequency domains.

The correlation function (CF) and the PSDF are related through the Fourier transform of each other, as shown by Lawson and Uhlenbeck [46]. When attempting to describe the frequency response of data obtained experimentally, the power functions of frequency are employed because some high-frequency components of the signal are larger than can be accounted for by a Gaussian distribution. There are many examples reported [32] where the PSDFs of radar systems vary as $1/(1 + f^2)$, $1/(1 + f^3)$, and $1/(1 + f^4)$. The functions having even powers of frequency have simple Fourier transforms, but the odd functions in frequency, $1/(1 + f^3)$, can only be approximated by series expansions in the time domain. This results in a complicated expression for the CF and is considered an excessive mathematical calculation at the present time.

It is common practice to consider two signals to decorrelate in time when the correlation function reaches e^{-1}, or 37% of its peak value. Table 3.16 gives decorrelation times calculated from the data, including both the slow-varying component and the fast-varying component. The long times are, of course, primarily due to the Gaussian or slow-varying component and are so marked on Figure 3.56.

Table 3.15
Measured Noncoherent Tree Spectral Dependencies

$W(f) = W_0 \exp(-af^2/f_t^2)$ (Gaussian)		$W(f) = \dfrac{A}{1 + \left(\dfrac{f}{f_c}\right)^n}$ (Lorentzian)		
Frequency (GHz)	Gaussian f_t (Hz)	Break Point (dB)	Wind	Lorenzian f_c	n
10	3.1	10	Low	—	3
10	3.1	10	High	9	3
16	2.25	12	High	—	3
16	2.25	12	High	16	3
35	2.8	15	Low	7	3
35	2.8	15	High	21	2.5
95	3.5	18	Low	6	2
95	3.5	18	High	35	2

Table 3.16
Decorrelation Time (e^{-1}) for Slow-Moving Components of Tree Spectrum

Band	τ (sec)
S	1.934
X	0.170
K_u	0.102
K_a	0.049
M	0.023

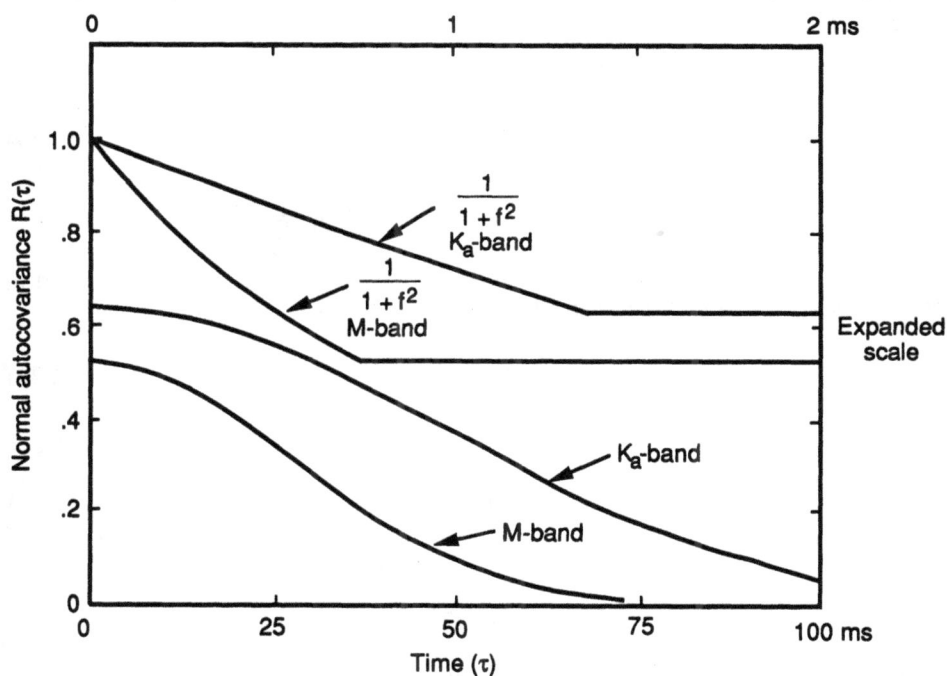

Figure 3.56 Autocovariance functions for noncoherent measurements of wind-blown trees.

Much of the spectral data collected by Georgia Tech from wind-blown trees has been processed with the dc and slowly varying components removed. In addition, the value of decorrelation time was specified as the time when the correlation function reached a value of 0.15, instead of e^{-1}. The following discussion concerns the high-frequency components.

Long, following Larson and Uhlenbeck [46], shows that the time domain and frequency domain are related through the Fourier transform pairs

$$R(t) = e^{-\alpha^2 t^2} \tag{3.16}$$

$$W(f) = \frac{2(\pi)^{1/2}}{\alpha} e^{-\pi^2 f^2/\alpha^2} \tag{3.17}$$

then

$$\frac{1}{2s_c^2} = \frac{\pi^2}{\alpha^2} \tag{3.18}$$

From Barlow's value of a $= 2.3 \times 10^{17}$ at 20 mph wind speed and radar frequency $f_t = 1$ GHz,

$$s_c = \frac{2}{\lambda} \left(\frac{c^2}{8a}\right)^{1/2}$$

$$= \frac{2}{0.3} \left(\frac{9 \times 10^{16}}{8 \times 2.3 \times 10^{17}}\right)^{1/2} \tag{3.19}$$

To keep the definition of the decorrelation time consistent at 0.15 to correspond to data collected by Georgia Tech at VHF, X-, K_u-, K_a-, and M-band, we can let

$$R(\tau) = 0.15 = e^{-\alpha^2 \tau^2} \tag{3.20}$$

Then

$$\ln(0.15) = -\alpha^2 \tau^2 \tag{3.21}$$

$$\tau^2 = \frac{-\ln(0.15)}{2\pi^2 s_c^2} \tag{3.22}$$

giving $t = 0.210$ sec at 1-GHz radar frequency.

Data were collected by Rivers [47] at Georgia Tech (1968) using a 140-MHz Doppler radar with pulse lengths varying from 0.1 to 0.5 μs. The data showed that the frequency spectra amplitude had disappeared after a 0.4-Hz spectrum spread. Since the dynamic range of the receiver was approximately 35 dB, we can assume that a value of 30 dB would correspond to Rivers' statement that the clutter spread has essentially disappeared. For the time correlation function,

$$R(\tau) = e^{-\alpha^2 \tau^2} \tag{3.23}$$

then

$$\alpha^2 \tau^2 = -\ln(0.15) \tag{3.24}$$

will be consistent with decorrelation times used at higher frequency radars, and $\alpha_\tau = 1/377$.

In the frequency domain, the normalized spectrum data reported by Rivers reduces

$$W(f) = \frac{2(\pi)^{1/2}}{\alpha} e^{-\pi^2 f^2 / \alpha^2} \tag{3.25}$$

to

$$0.001 = e^{-\pi^2 (0.04)^2 / \alpha^2} \tag{3.26}$$

This will give a value of $\alpha = 0.4781$ and of $\tau = 2.88$ sec, from the time-equation $\alpha \tau = 1.377$. This is the time required for the autocorrelation function to decay from a value of 1 at time zero to a value of 0.15. This value of decorrelation time is shown in Figure 3.57 along with other microwave and millimeter-wave measurements.

The decorrelation times at X-, K_u-, K_a-, and M-band were extrapolated to 14 mph directly from data collected by Georgia Tech with noncoherent radars having pulse lengths of 0.1 μs, narrow antenna beams of approximately 1°, and targets at ranges of about 1 km. The Georgia Tech data are shown in Figure 3.58.

Amplitude Fluctuations. As indicated in Section 1.5.1, temporal amplitude fluctuations are those variations that occur within a single radar cell with passing time and are described by amplitude statistics. Data have been obtained on the amplitude variations of trees and grass at millimeter wavelengths and are summarized in Table 3.17, which lists the measured standard deviations of returns from several types of trees and grass at 10, 16, 35, and 95 GHz. For the 10-GHz data the distribution standard deviations are 3 to 4 dB, which is characteristic of a Rayleigh distribution, but at 95 GHz the standard deviation is larger and the distributions more closely approximate a lognormal as opposed to a Rayleigh distribution. Figure 3.59 gives a

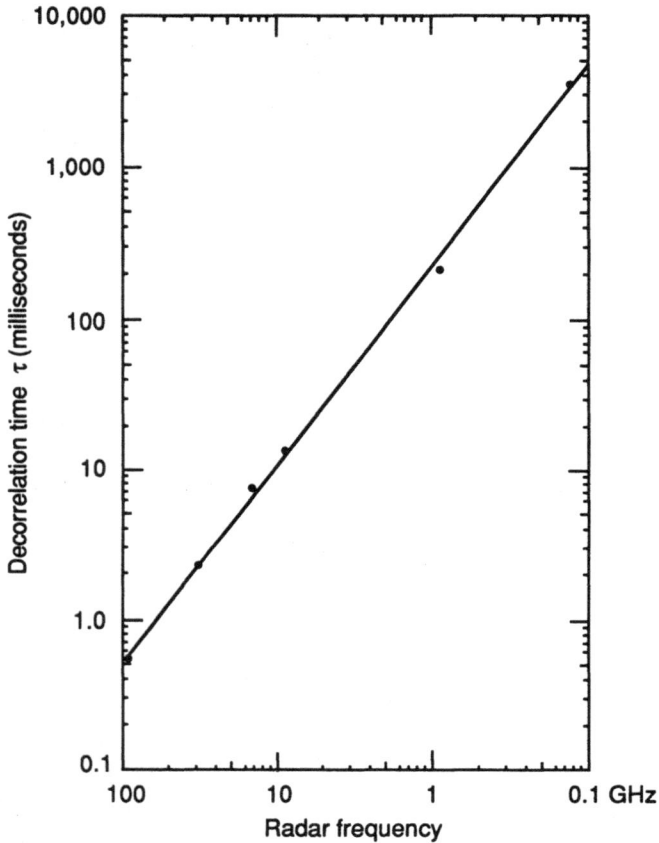

Figure 3.57 Decorrelation time versus frequency for wind-blown trees based on microwave data.

temporal amplitude distribution for deciduous trees at 35 GHz, while Figure 3.60 gives the distribution for the same clutter cell for 95 GHz. As can be seen, the 35-GHz data are more closely approximated by a Rayleigh distribution, while the 95-GHz data more closely resemble a lognormal distribution. The data in Table 3.17 and Figures 3.60 and 3.61 are narrowband data. Wideband processing would be expected to narrow the amplitude variations.

Spatial Variations. Spatial variations of clutter are important whenever a radar scans the land surface looking for specific targets of interest. Since the terrain is not generally uniform, variations in the backscattered signal to the radar occur. The typical

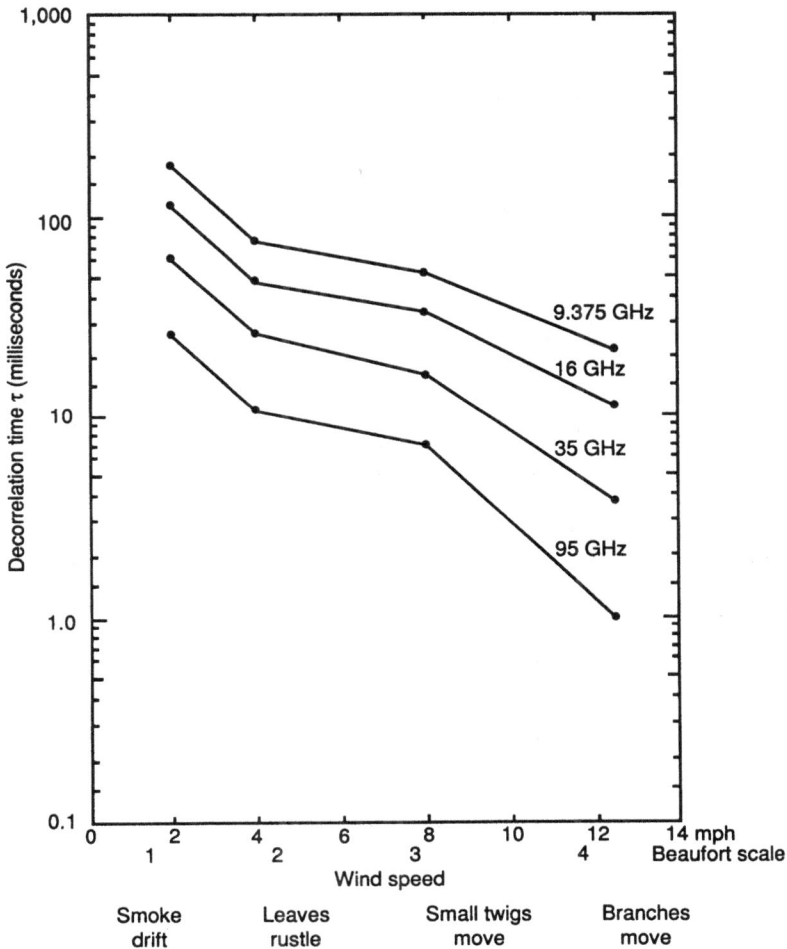

Figure 3.58 Decorrelation time versus frequency for wind-blown trees based on Georgia Tech data. (From Currie, Dyer, and Hayes [48].)

terrain consists of areas with relatively uniform properties adjacent to other areas with different features, but also relatively uniform, such as a field next to a forest. Thus, it is important to consider not only variations within a given terrain type but also what happens when the radar beam scans across different types of terrain. Generally, at microwave frequencies, the spatial distributions within uniform areas are

Table 3.17

Standard Deviations for Foliage Versus Frequency (From Currie, Dyer, and Hayes [48])

Clutter Type	Polarization	*Average Value of Standard Deviation (GHz)*			
		9.5	16.5	35	95
Deciduous Trees	Vertical	3.9	—	4.7	—
Summer	Horizontal	4.0	—	4.0	5.4
	Average	4.0	—	4.4	6.6
Deciduous Trees	Vertical	3.9	4.2	4.4	6.4
Fall	Horizontal	3.9	4.3	4.3	5.3
	Average	3.9	4.2	4.3	5.0
Pine Trees	Vertical	3.5	3.7	3.7	6.8
	Horizontal	3.3	3.8	4.2	6.3
	Average	3.4	3.7	3.9	6.5
Mixed Trees	Vertical	4.3	—	4.0	—
Summer	Horizontal	4.6	—	4.2	—
	Average	4.4	—	4.1	—
Mixed Trees	Vertical	4.1	4.1	4.7	6.3
Fall	Horizontal	4.5	4.3	4.6	5.0
Tall Grassy Field	Vertical	1.0	1.2	1.3	—
	Horizontal	1.0	1.2	1.3	—
	Average	1.3	1.2	1.4	2.0
Rocky Area	Vertical	1.1	2.2	1.8	1.7
	Horizontal	1.2	1.7	1.7	1.7
	Average	1.1	1.9	1.8	1.7
10″ Corner Reflector (located in grassy field)		1.0	1.0	1.2	1.2

Rayleigh distributed except for elevation angles less than 5° [49]. Some data on the spatial distributions of various types of terrain have been measured at 95 GHz and are listed in Table 3.18, which gives the measured standard deviations of uniform clutter areas at three depression angles, from Lane [50]. When the statistics of the radar return from a number of terrain types are considered, the distribution appears more nearly lognormal, and the standard deviation is greater. Figure 3.61 gives the spatial distribution of a number of terrain backgrounds for 95 GHz, from Lane. The scale shows results in a straight line if the data are lognormally distributed. As can be seen, a straight line nearly fits the data.

The measured standard deviations in Table 3.18 are relatively small because the radar that was used averaged the data over a 256-MHz bandwidth, thus greatly

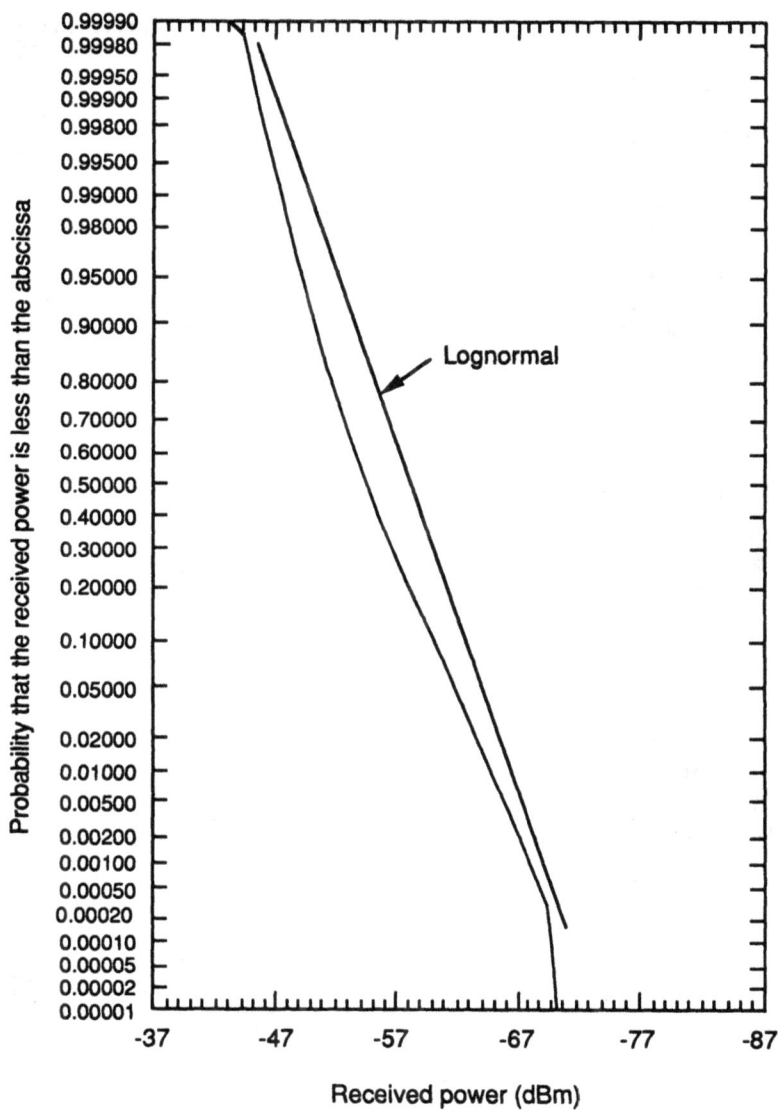

Figure 3.59 Cumulative probability distribution of the received power from deciduous trees at 35 GHz. (From Currie, Dyer, and Hayes [48].)

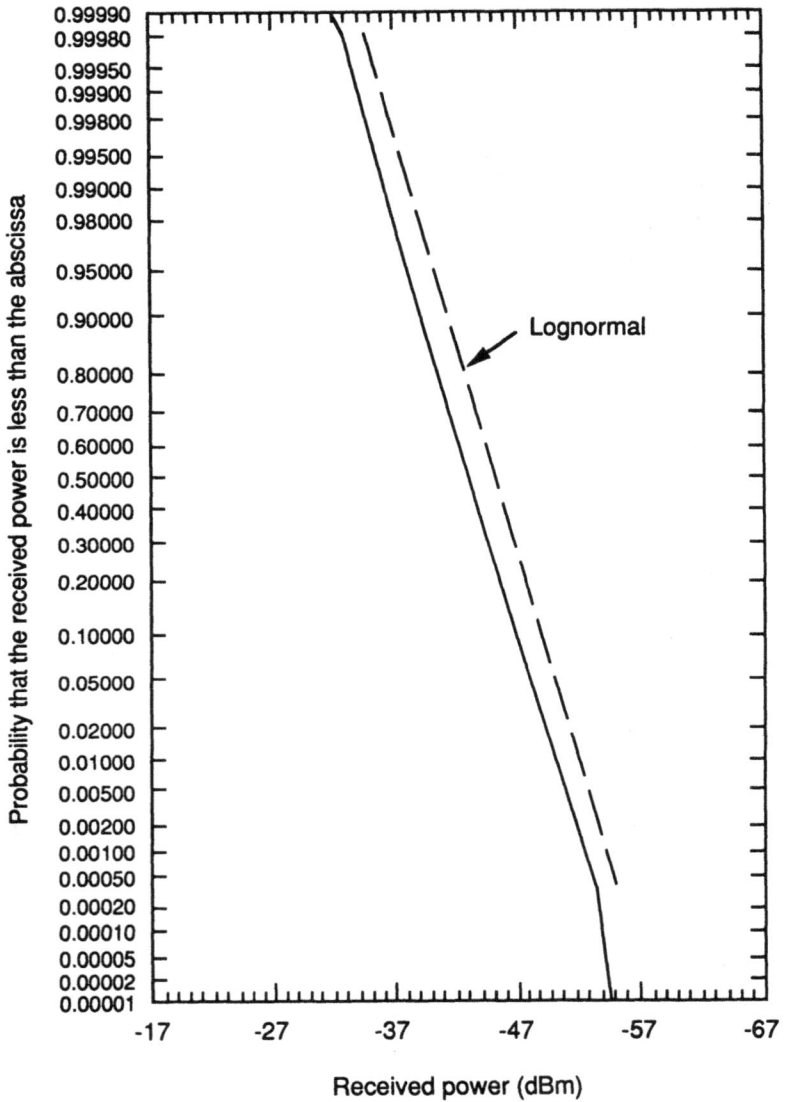

Figure 3.60 Cumulative probability distribution of the received power from deciduous trees at 95 GHz. (From Currie, Dyer, and Hayes [48].)

Table 3.18
Standard Deviations for Uniform Clutter at Three Depression Angles (From Lane [31])

Clutter Type	Depression Angle (°)	Standard Deviation (dB)
Barley	12	1.1
	30	2.1
Coniferous Trees	12	1.4
	30	1.7
	75	2.25
Corn	12	1.0
	30	1.7
	75	2.45
Deciduous Trees	12	2.0
	30	2.0
Plowed Field	12	1.0
	30	1.5
	75	2.5
Grass	12	0.9
	30	1.9
	75	3.3

Note:
These data were averaged over 256 MHz with an effective pulse bandwidth ($1/\tau$) of 5 MHz. The number of independent samples averaged was thus $B_{rf}/2B_{vid} = 256/10 = 25.6$. The effective standard deviations of the original data are thus those given $\sqrt{25.6} = 5.05$, which are definitely non-Rayleigh.

reducing the variations in the returns. As the note at the bottom of the table indicates, the effective number of independent samples is 25.6, yielding a narrowing of the distributions of $\sqrt{25.6} = 5.05$. If wideband averaging had not been used, then the variations would have been a factor of 5 greater.

The amount of frequency averaging to decorrelate clutter is a function of the radar resolution cell and the number of scatterers in the beam. Figure 3.62 gives the measured amount of bandwidth required to decorrelate the spatial variation of snow-covered ground at 35 GHz as a function of footprint size (beam-limited scenario with changing cross-range dimension with grazing angle) [54]. The smaller the footprint size, the more bandwidth is required to decorrelate the snow clutter. Thus, wideband frequency averaging can be seen to be very helpful in reducing variations in uniform terrain, but small beamwidths require large amounts of frequency agility.

Lognormal scale
95 GHz, VV, 149, 151 pixels

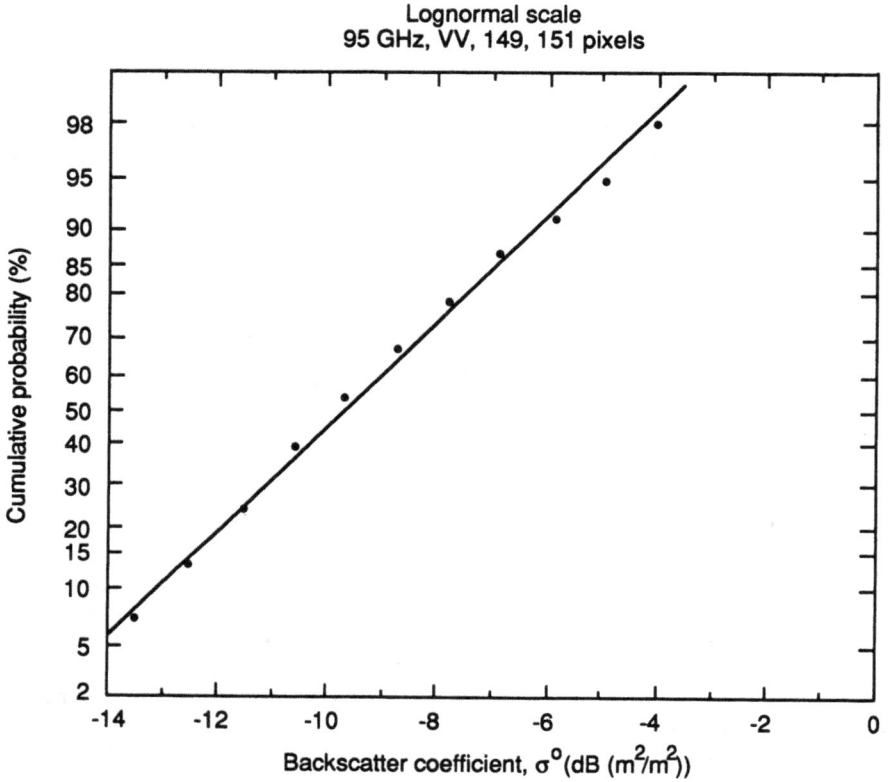

Figure 3.61 Spatial distribution for a number of different clutter types at 95 GHz. (From Lane [50].)

3.3.2 Sea Clutter Characteristics

3.3.2.1 Definitions

The surface of the sea moves with time and is strongly affected by interaction with the wind and tides. Thus, some definitions that are unique to the problem of describing the sea surface need to be further elaborated. Table 3.19 summarizes these definitions, which include sea state, average wave height, significant wave height, peak wave height, wind/wave direction, and fetch. All of these terms are associated with attempts to describe the surface of the sea at a given time. They are important because radar reflectivity data are often described in relationship to one or more of these parameters, and many sea clutter models require wave height as an input variable. More detailed information on the definition of these parameters is available in Long [4].

35 GHz, VV polarization, log receiver
T-102; T-128; T-131

(a)

95 GHz, VV polarization, log receiver
T-102; T-128; T-131

(b)

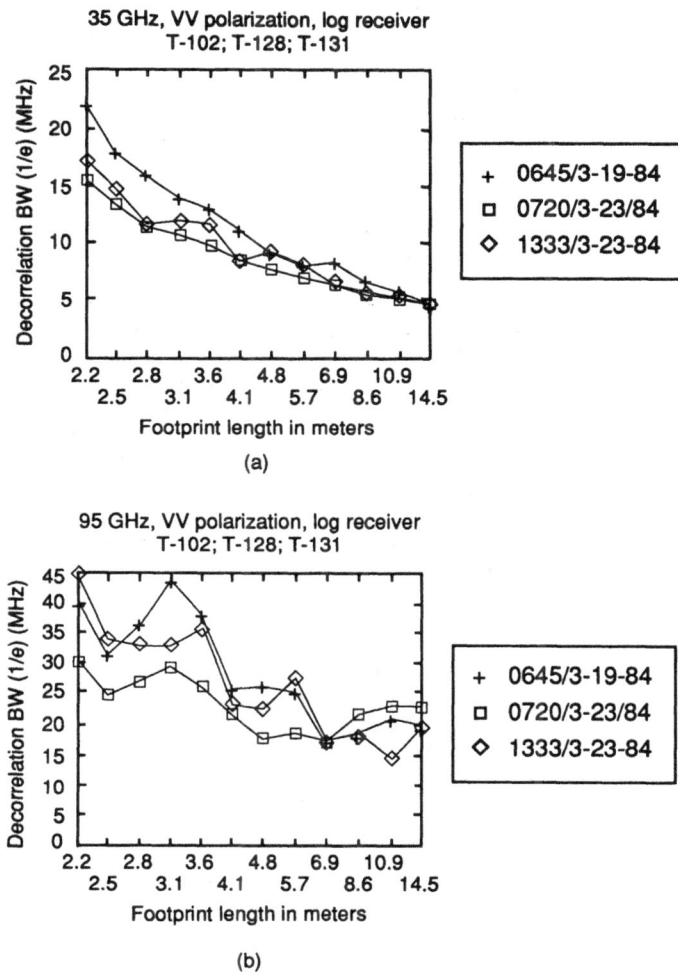

Figure 3.62 Decorrelation bandwidth for snow-covered ground versus pulse extent in meters at 35 GHz. (From Currie et al., © 1988 by IEEE.)

3.3.2.2 General Dependencies

To date, only a small number of measurements have been performed on the reflectivity of the sea at millimeter wavelengths in contrast to the large body of data that exists at microwave frequencies. Thus, the best way to discuss millimeter-wave reflectivity characteristics is to compare the little data that exist at millimeter wavelengths to those at lower frequencies. This is done in Table 3.20. For microwave

Table 3.19
Sea Clutter-Related Definitions

Parameter	Definition
Sea State	A descriptive scale for the roughness of the sea based on a given trough-to-crest height of the highest one-third of the waves. See Long [4, Sec. 3.3.1.3.-10] for more information.
Average Wave Height	The true mean peak-to-trough height of ocean waves at a given time.
Significant Wave Height	The peak-to-trough height of the highest one-third of the waves. It is thought to be the height given by an observer.
Peak Wave Height	The peak-to-trough height of the highest one-tenth of the waves. It is approximately twice the average wave height.
Wind/Wave Direction	The direction of the wind and waves relative to the radar line of sight.
Fetch	The interaction distance between the wind and water which leads to a given wave height for a given wind speed.
Fully Developed Sea	A sea for which enough time has passed for the waves to reach their maximum value for a given fetch and wind speed.

Table 3.20
Sea Clutter Radar Reflectivity Characteristics

Characteristic Polarization Dependence	Value	
	Microwave	Millimeter Wave
Average Values		
Low Grazing Angles	$\sigma^0_{HH} < \sigma^0_{VV}$	$\sigma^0_{HH} > \sigma^0_{VV}$
Higher Grazing Angles	$\sigma^0_{HH} \approx \sigma^0_{VV}$	$\sigma^0_{HH} \approx \sigma^0_{VV}$
Standard Deviations		
All Grazing Angles	$\sigma_{HH} > s_{VV}$	$\sigma_{HH} \approx \sigma_{VV}$
Wind/Wave Dependence		
High Grazing Angles	$\sigma^0_{UPWIND} \approx \sigma^0_{CROSSWIND} \approx \sigma^0_{DOWNWIND}$	
All Other Grazing Angles	$\sigma^0_{UPWIND} > \sigma^0_{CROSSWIND} > \sigma^0_{DOWNWIND}$	
Frequency Dependence		
Average Values	$\sigma^0_{95\,GHz} < \sigma^0_{10\,GHz} < \sigma^0_{35\,GHz}$	
Standard Deviations	$\sigma_{10\,GHz} < \sigma_{35\,GHz} < \sigma_{95\,GHz}$	

frequencies, the reflectivity at low-to-moderate elevation angles is dependent on the wind speed and direction and the wave height and direction. At microwave wavelengths for a "fully developed" sea, where the waves have reached their maximum height for the wind speed and are in the same direction as the wind, the reflectivity is proportional to the wave height, is a maximum in the upwind/upwave direction, and is a minimum in the downwind/downwave direction. This trend appears to also hold at millimeter wavelengths. The average value of hh polarization is lower than vv polarization at microwaves for low wavelengths, while this effect appears to be reversed at millimeter wavelengths. Finally, the average value of the return at 10 GHz is less than that at 35 GHz, but greater than that at 95 GHz.

The standard deviations measured for millimeter-wave sea return are also different from those measured at microwaves. At microwave lengths the standard deviation for hh polarized returns are almost always higher than those for vv returns except for very rough seas. This trend is reversed at 95 GHz, where the standard deviations of vv polarized returns have often been found to be larger than hh polarized returns.

The dependence of the reflectivity on wind/wave direction seems to be independent of the wavelength up through 95 GHz. That is, the return from the sea is always greatest in the upwind/upwave direction and the least in the downwind/ downwave direction, except at depression angles near nadir, where the return is approximately independent of wind/wave direction. When the wind and wave direction are not the same, the backscatter dependence is variable and difficult to predict.

3.3.2.3 Amplitude Characteristics

Average Values. Figure 3.63 gives measured data on sea return at 9.4, 35, and 95 GHz for vv and hh polarizations compared with the Georgia Tech millimeter-wave sea clutter model predictions discussed in the next section. These data are for upwind conditions and moderate sea states (2- to 3-ft average wave height), and thus represent the highest backscatter conditions. The vv data are higher than the hh data at 9.4 GHz, while they are approximately equal at 35 and 95 GHz.

Figure 3.64 is a scatter diagram that compares the reflectivity of the sea as measured at 100 GHz versus the reflectivity measured simultaneously at 10 GHz. As can be seen from the figure, the return at 10 GHz is larger than that at 100 GHz most of the time. Figure 3.65 illustrates the change in polarization dependence as a function of frequency. The figure is another scatter diagram that compares σ_{vv}^0 with σ_{hh}^0 at 35 GHz and 95 GHz, respectively. For the 35-GHz data, σ_{vv}^0 is generally larger than σ_{hh}^0, while at 95 GHz the two returns are approximately equal, with a slight edge for σ_{hh}^0 over σ_{vv}^0.

To date, no data have been published for cross-polarized returns from the sea at millimeter wavelengths (i.e., hv or vh), but it would be expected that these data

(a)

Vertical polarization

(b)

Horizontal polarization

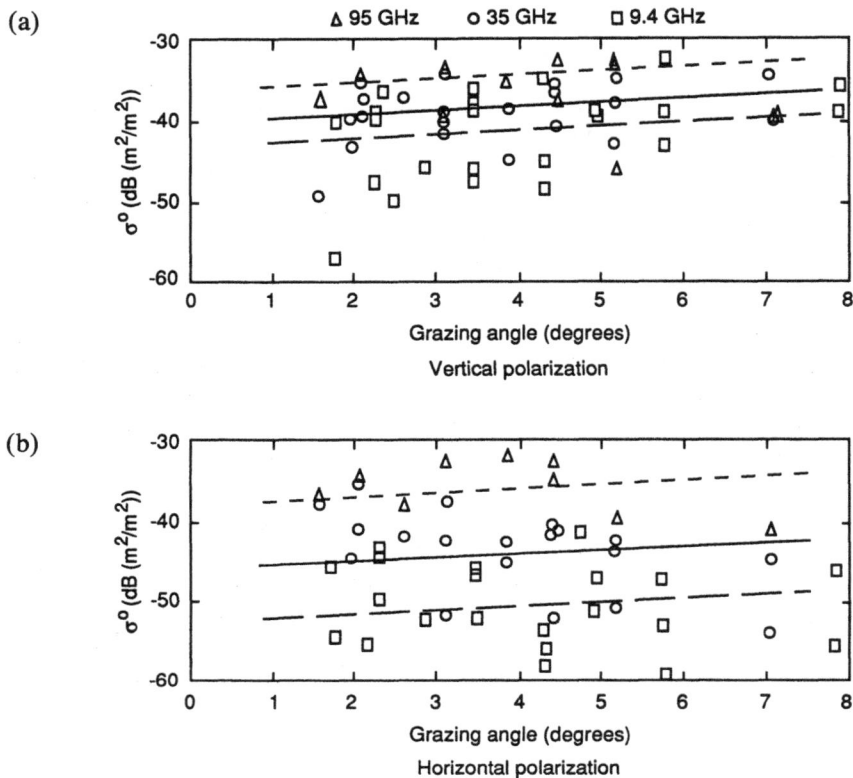

Figure 3.63 Comparison of sea return at 9.4, 35, and 95 GHz for (a) vv polarization and (b) hh polarization.

would generally be a few decibels less than the parallel polarized returns, as is the case for microwave returns. Also, no known circularly polarized data are available but a difference of only a few decibels would be expected from the linearly polarized data.

Average Value Model (Adapted from Horst and Perry [54]). As discussed in Chapter 2, the surface of the sea does lend itself to theoretical modeling under some conditions and at some wavelengths [55]. At millimeter wavelengths, however, even very tiny structures become important, and it is no longer possible to concentrate on just gravity waves and ignore surface ripples and similar fine structure effects. As a result of these difficulties, an empirical/analytical model was developed by researchers at Georgia Tech [56]. This is by no means the only such model, and others can be found in the references.

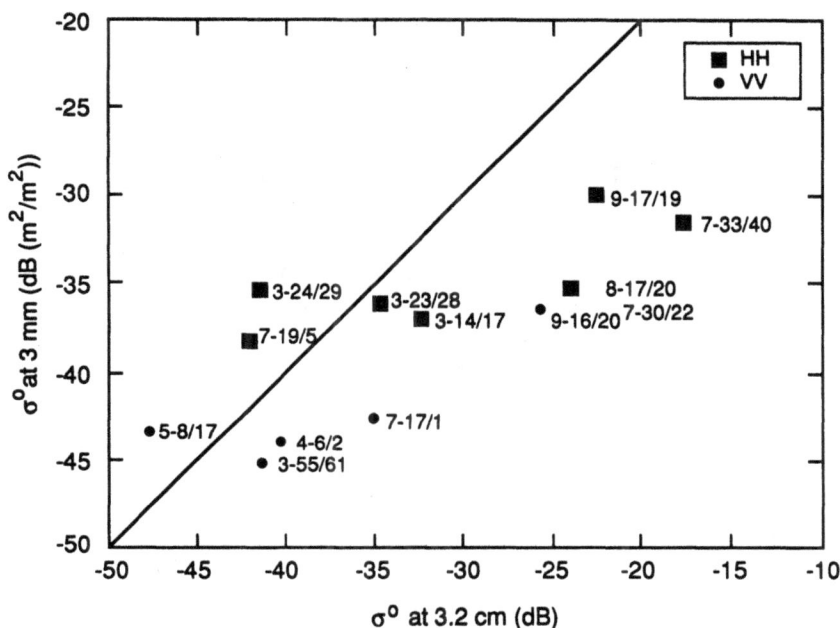

Figure 3.64 Scatter diagram of sea return at 95 GHz versus the return at 10 GHz. (From Rivers [52].)

The problem of developing a model for sea backscatter is a complicated one at best, since it is dependent upon radar frequency, sea state, incidence angle, wind/sea direction, polarization, and perhaps other factors. Researchers at Georgia Tech designed an analytical model to describe sea clutter backscatter using empirical constants derived from measured data.

For ease of reference, Tables 3.21 and 3.22 present a list of the variables in the Georgia Tech sea clutter model and a summary of the basic equations of the model for the frequency range 10 to 100 GHz. The equations in the first section of Table 3.22 serve to define the incidence angle, taking into account both the curvature of the earth and the possibility of ducting or enhanced propagation conditions. The propagation of an electromagnetic wave in a surface duct of height h_d can be effectively modeled by clamping the incidence angle in this manner. The equations in the second and third sections will be explained in detail below, while the equations in the last section provide a means of expressing the total clutter cross section seen by the radar as the product of the clutter cross section per unit area times the radar clutter cell size.

The clutter model calculates σ^0 as the product of three variables: for multipath interference, sea direction, and wind speed [57]. Each of these factors is, in turn, a

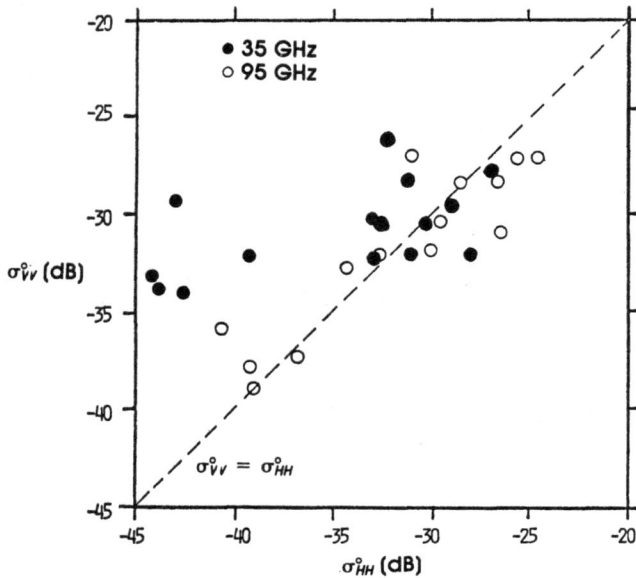

Figure 3.65 Scatter diagram of sea return measured simultaneously for vv and hh polarization at 3. and 95 GHz versus the return at 10 GHz. (From Trebits, Currie, and Dyer [53].)

function of the appropriate independent variables. The factor describing the multipatl interference effect, between the direct and scattered fields from the surface for horizontal polarization, is derived from forward scatter theory assuming a Gaussian distribution of surface height with standard deviation σ_h. A surface reflection coefficien of -1 is assumed for all angles of incidence. A roughness parameter σ_r may be defined as

$$\sigma_r = \frac{4\pi \sin \alpha \, \sigma_h}{\lambda} = \frac{2\sqrt{2\pi} \, \alpha \, h_{av}}{\lambda} \qquad (3.27$$

where λ is radar wavelength, σ_h is the rms surface roughness, h_{av} is average wave height, and the angle of incidence α is small enough that $\sin \alpha \approx \alpha$.

The interference term A_1 is then given by

$$A_i = \sigma_r^4/(1 + \sigma_r^4) \qquad (3.28$$

The behavior of σ^0 versus incidence angle falls into two distinct regions: a low grazing angle region in which σ^0 is a strong function of angle, and a plateau regior

Table 3.21
Georgia Tech Millimeter-Wave Sea Clutter Model Variables

Symbol	Definition	Restrictions
h_a	Radar antenna height	—
R	Range	—
A_e	Effective earth radius	—
λ	Radar wavelength	0.003m–0.3m
		(0.1–1 ft)
h_d	Duct height constant	—
α	Incidence angle	0.1°–10°
h_{av}	Average wave height	0m–4m (0–13 ft)
Φ	Angle between boresight and upwind	0°–180°
τ	Pulsewidth	50 ns–2 μs
Θ_a	3-dB azimuth beamwidth (one-way)	—
c	Speed of light	—
σ^0	Average clutter cross section per unit area	—
A_c	Area of radar resolution cell	—
σ_c	Average clutter cross section	—
A_i	Interference factor	—
A_u	Upwind-downwind factor	—
A_w	Wind speed factor	—

in which σ^0 is approximately independent of angle. These two regions are separated by a critical angle a_c, which is generally considered to be proportional to λ and $1/h_{av}$:

$$\alpha_c = \frac{\lambda}{K h_{av}} \tag{3.29}$$

where K is a proportionality constant.

No data could be found to define the critical angle at millimeter wavelengths, since α_c would probably be well below the practical angles possible in a measurement program. Therefore, only σ^0 data clearly from the plateau region were used in determining parametric dependencies for the model.

The sea direction term is based on upwind-downwind ratio data, although the reference for the aspect angle Φ should be the sea wave propagation vector rather than the wind vector. The up-down ratio is dependent on incidence angle α, increasing as α decreases, but approaching a finite value at $\alpha = 0$. Since there are insufficient data to determine clearly the functional form of the dependence, a simple

Table 3.22
Georgia Tech Millimeter-Wave Sea Clutter Model Equations

h_a, h_d, λ, h_{av}, (m)

R, A_e (km)

V_w (m/s)

σ_r = surface roughness

$\alpha' = h_a/1000R - R/2A_e$

$$\alpha = \left[\alpha'^2 + \left(\frac{l}{4h_d}\right)^2 \right]^{1/2}$$

$\sigma_r = 2\sqrt{2\pi}\alpha h_{av}/(\lambda + 0.015)$

$A_i = \sigma_r^4/(+ \sigma_r^4)$

$A_u = \exp[0.25 \cos \Phi (1 - 2.8\alpha) \lambda^{-0.33}]$

$q_w = 1.93\lambda^{-0.04}$

$V_w = 8.67 h_{av}^{0.4}$

$A_w = [1.94 V_w/(1 + V_w/15.4)]^{q_w}$

$\sigma_{hh}^0 = 10 \log (5.78 \times 10^{-6} \, \alpha^{0.547} \, A_i A_u A_w$ (dB)

$\sigma_{vv}^0 = \sigma_{hh}^0 - 1.38 \ln(h_{av}) + 3.43 \ln(\lambda) + 1.31 \ln(\alpha) + 18.55$ (dB)

$$A_c = 10 \log \frac{1000R\theta_a c\tau}{2\sqrt{2}}$$

$\sigma_c = \sigma^0 + A_c$ (dB)

circular function, $\cos \Phi$, is chosen, where Φ is the angle between antenna boresight and the upwind direction. The upwind-downwind term A_u is then given by

$$A_u = \exp[0.156(1 - 2.8\alpha)(0.5 \cos \Phi)c_1\lambda^c] \qquad (3.30$$

For the purpose of determining the wavelength coefficient and exponent, the relation between the upwind and downwind backscatter is modeled by the equation

$$\frac{\sigma_{up}^0/\sigma_{down}^0}{1 - 2.8\alpha} = c_1\lambda^{c_2} \qquad (3.31$$

Due to a lack of upwind, downwind, and crosswind data for higher frequencies the following assumption is also made:

$$(\sigma_{up}^0/\sigma_{down}^0)_{MMW} = 2(\sigma_{up}^0/\sigma_{down}^0)_{Microwave} \qquad (3.32$$

With the additional upwind-crosswind data, the wind direction term A_u becomes

$$A_u = \exp[0.25 \cos \Phi(1 - 2.8\alpha)\lambda^{-0.33}] \qquad (3.33)$$

The dependence of σ^0 on sea state has been shown to be more a function of wind speed than of wave height. The apparent saturation of σ^0 with increasing wind speed, V_w, is accounted for using a wind speed term, A_w, which is computed in two steps under the assumption that

$$\sigma^0 = (\text{constant})(V_w)^{c_1\lambda^{c_2}} \qquad (3.34)$$

First, wind speed exponents β are found using σ_{hh}^0 grouped by incidence angle, wavelength, and source. This provides nine β values. Then the LSF to

$$\log_{10}\beta = \log_{10}c_1 + c_2 \log_{10}\lambda \qquad (3.35)$$

is computed using these β values and their associated wavelengths to produce the result

$$\sigma^0 = (\text{constant})V_w^{1.93\lambda^{-0.04}} \qquad (3.36)$$

The small exponent on λ indicates little wavelength dependence in the wind speed term.

Using the three model components thus computed, the reflectivity σ^0 is found by making a fit to

$$10^{\sigma_{hh}^0/10}/\alpha^{0.4} A_i A_u A_w = c_1\lambda^{c_2} \qquad (3.37)$$

Using the least squares approach on

$$\log_{10}(10^{\sigma_{hh}^0/10}/\alpha^{0.4} A_i A_u A_w) = \log_{10}c_1 + c_2 \log_{10}\lambda \qquad (3.38)$$

results in an expression for σ_{hh}^0:

$$\sigma_{hh}^0 = 10 \log_{10}[3.5 \times 10^{-6}\lambda^{0.01}\alpha^{0.4}A_i A_u A_w] \qquad (3.39)$$

The wavelength exponent is approximately zero, and for simplification purposes it is removed. Without λ in the equation, a new angle dependence can be determined by making an LSF to

$$\log_{10}(10^{\sigma_{hh}^0/10}/A_i A_u A_w) = \log_{10}c_1 + c_2 \log_{10}\alpha \qquad (3.40)$$

This produces the final result for σ_{hh}^0:

$$\sigma_{hh}^0 = 10 \log_{10}[5.78 \times 10^{-6}\alpha^{0.54}A_i A_u A_w] \qquad (3.41)$$

The correction factor for vertical polarization is assumed to follow the form of a linear fit to sea state (average wave height), wavelength, and incidence angle. Only data for which σ_{hh}^0 and σ_{vv}^0 exist at approximately the same sea state, wavelength and angle are used to obtain the coefficients:

$$\sigma_{vv}^0 - \sigma_{hh}^0 = -1.38 \ln (h_{av}) + 3.43 \ln (\lambda) + 1.31 \ln (\alpha) + 18.55 \quad (3.42)$$

Figures 3.66 and 3.67 are examples of the predictions of the Georgia Tech sea clutter model. They present σ_{hh}^0 and σ_{vv}^0 predictions as a function of frequency from 10 to 100 GHz for average wave heights of 0.15 m, 0.6 m, 1.5 m, and 3.5 m, at incidence angle of 1.2°, and crosswind direction. In an effort to show more points for comparison, data from five sources, including σ^0 data at incidence angles of 1.0° and 1.4° and from an upwind direction were plotted with the predictions. The data should be just above the knee of the curve into the plateau region, so the variation in incidence angle should have negligible effect on the comparison. The difference between upwind and crosswind could account for some of the differences between measured and predicted σ^0 values, but the wind direction contribution should be less than the spread of the data within one experiment.

Figures 3.66 and 3.67 illustrate that the σ^0 model equations from 10 to 100 GHz do provide a reasonable fit to the data, considering the scarcity of available measured data points and the spread of the data. The rms error between measured

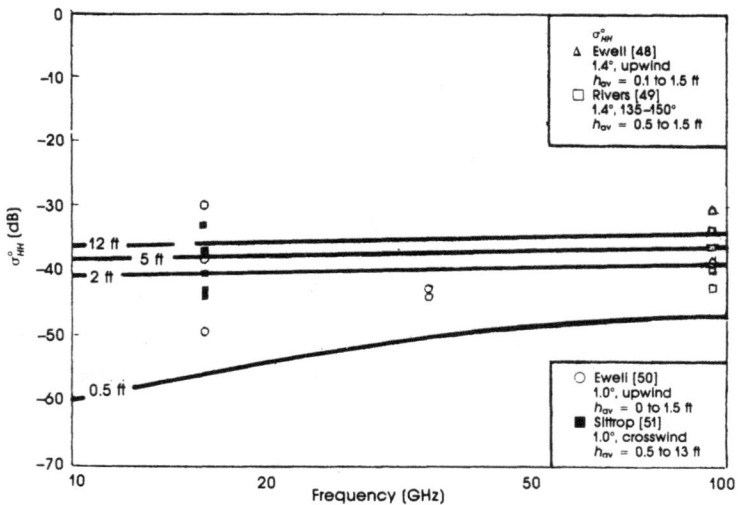

Figure 3.66 Sea return predictions by the Georgia Tech Model compared with measured data from 1 to 100 GHz, horizontal polarization. (From Horst, Dyer, and Tuley, © 1978 by IEEE.)

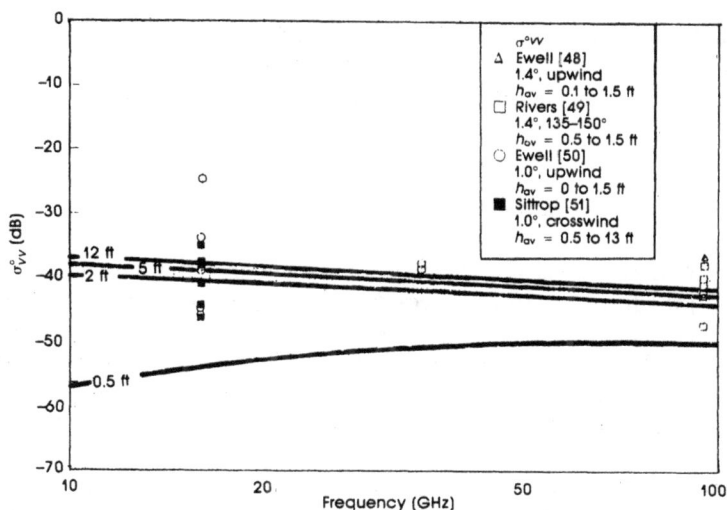

Figure 3.67 Sea return predictions by the Georgia Tech Model compared with measured data from 10 to 100 GHz, vertical polarization. (From Horst, Dyer, and Tuley, © 1978 by IEEE.)

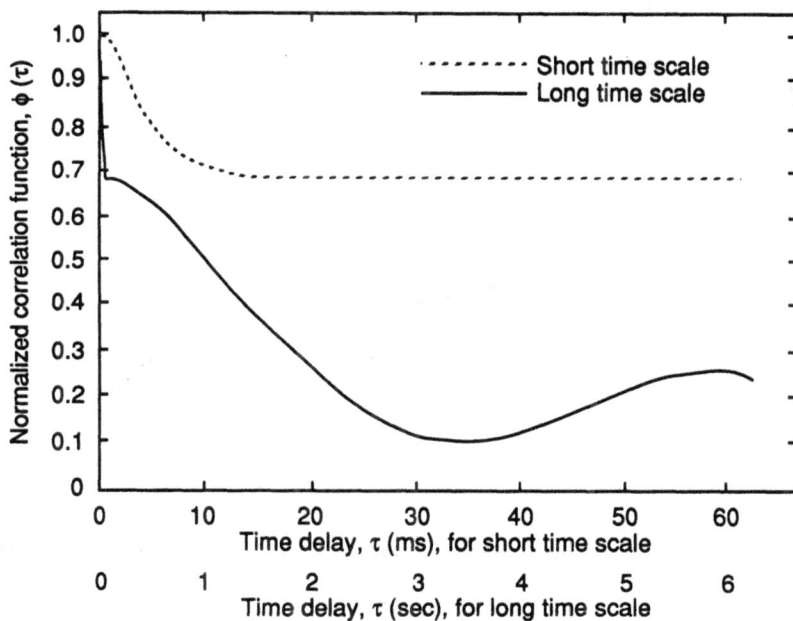

Figure 3.68 Illustration of the characteristics measured on correlation functions of sea clutter. (From Long, © 1984 by Artech House, Inc.)

and predicted σ^0 values was 3.7 dB for horizontal polarization and 5.9 dB for vertical polarization. Since the model equations were designed using all of the available data, these rms errors reflect the quality and consistency of the data base more than the accuracy of the model.

3.3.2.4 Sea Temporal Characteristics

Time correlation functions (CFs) are presented in the literature by Long [4] and Hayes et al. [58] from experiments conducted along the east coast of the U.S. employing X-band radars. PSDFs have been reported by Sittrop [59] from experiments conducted along the west coast of northern Europe employing an X-band radar. Some of the data are presented in Figures 3.68 through 3.70, which show details in the low-frequency spectrum (less than 20 Hz) and both long and short times in the correlation domain.

Trebits and Perry [60] have reported a large amount of data of PSDFs time CFs for sea clutter from a four-frequency measurement program collected during the late 1970s. One complete set of data considered typical of upwave/crosswind in sea state 2/3 have been modeled here in the same manner as that done for the X-band data reported by Long and Sittrop. The examples are shown in Figures 3.71 through

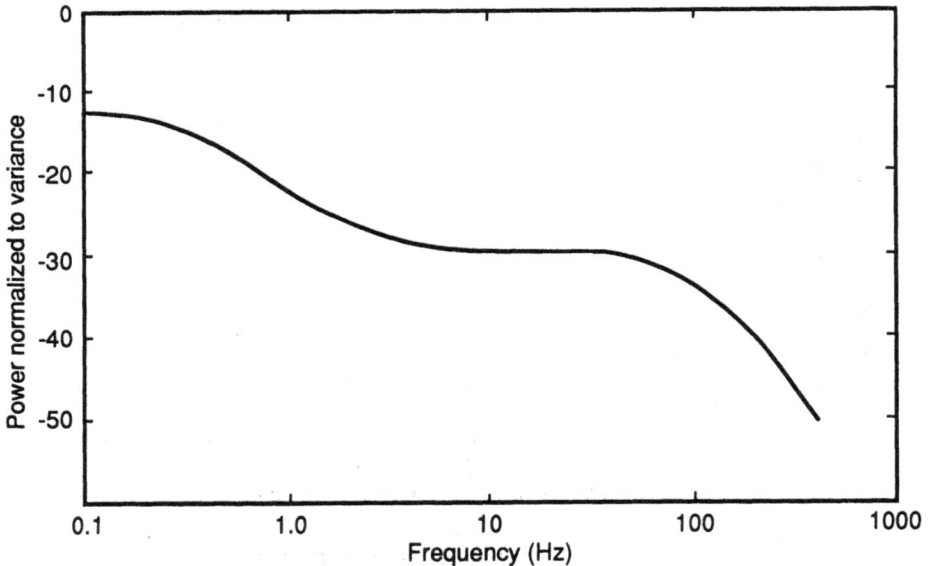

Figure 3.69 X-band sea clutter fluctuation spectrum, horizontal polarization. (From Sittrop [59].)

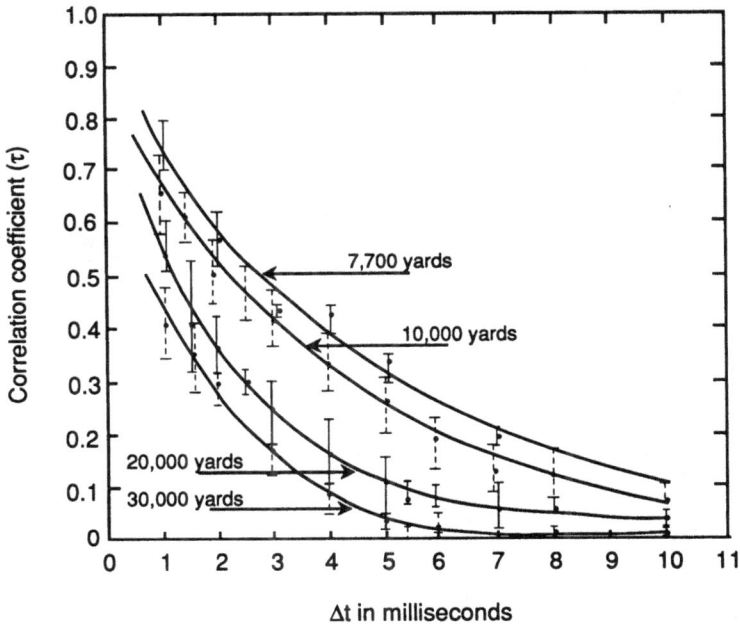

Figure 3.70 Time correlation of sea return, statistical variations. (From Hayes [58].)

3.74. The Georgia Tech data processing does not show fine grain details of the PSDF at frequencies below 25 Hz, nor fine grain details of the CF at times less than 10 ms. These limitations are no more restrictive than the fact that Long only reported CF and no PSDF, and Sittrop reported only PSDF and no CF. The concept and the theory that the long-time Gaussian portion of the CF will transform into a low frequency Gaussian portion of the PSDF and that the high-frequency Lorentzian portion of the PSDF will transform into the short-time exponential portion of the CF appear to be valid from the X-, K_u-, K_a-, and M-band sea backscatter data, as were true for tree backscatter data.

As discussed in Section 3.3.1.2 on ground clutter spectra, the Fourier transform F relating the PSDF in the frequency domain to the CF in the time domain can be determined when the functions are well behaved and the Fourier functions exist. Sea clutter has been divided into three components by Long [4] and are

1. Fast movers resulting from spray and white caps.
2. Slow movers due to passing surface wave fronts.
3. Very slow moving swells or gravity waves in the sea surface.

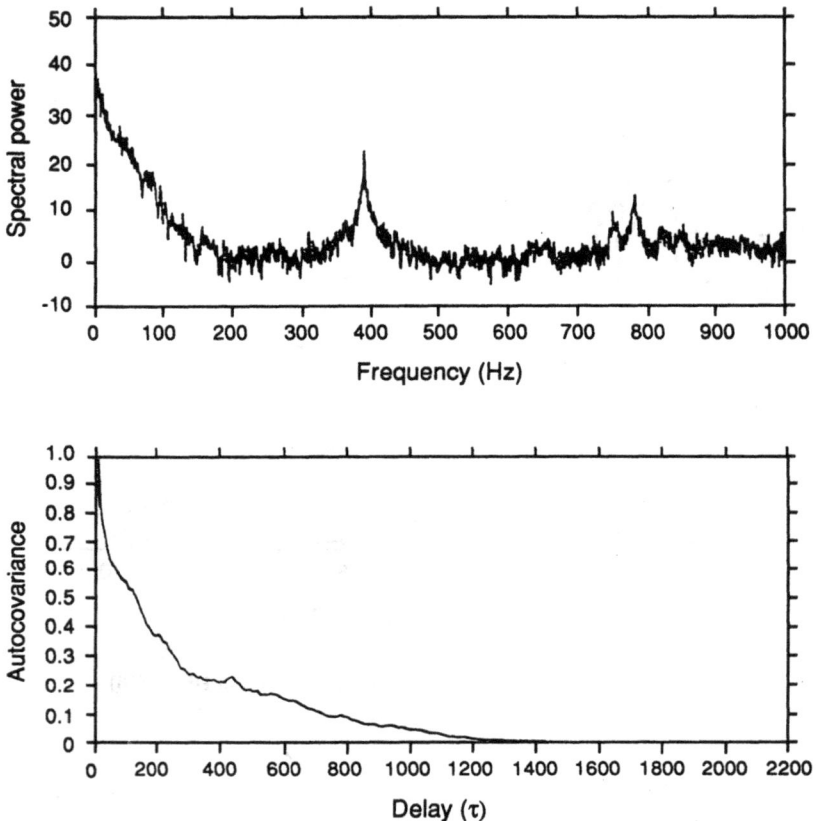

Figure 3.71 Sea return measured spectrum and autocovariance for 10 GHz, hh polarization. (From Trebits and Perry [60].)

Table 3.23 presents calculations of the two parts of the autocorrelation function based on the data from Trebits and Perry.

Two items are to be noted in the Trebits and Perry data:

1. There is 400-Hz power supply noise in the X-band data.
2. No heavy gravity waves were recorded, most likely a result of the measurement location in the Gulf of Mexico, where the depth is typically shallow.

Assuming that the three components in the sea clutter spectrum are independent, then the Fourier transforms of each part can be determined separately and added independently. From measured data, the general form of the various parts of clutter

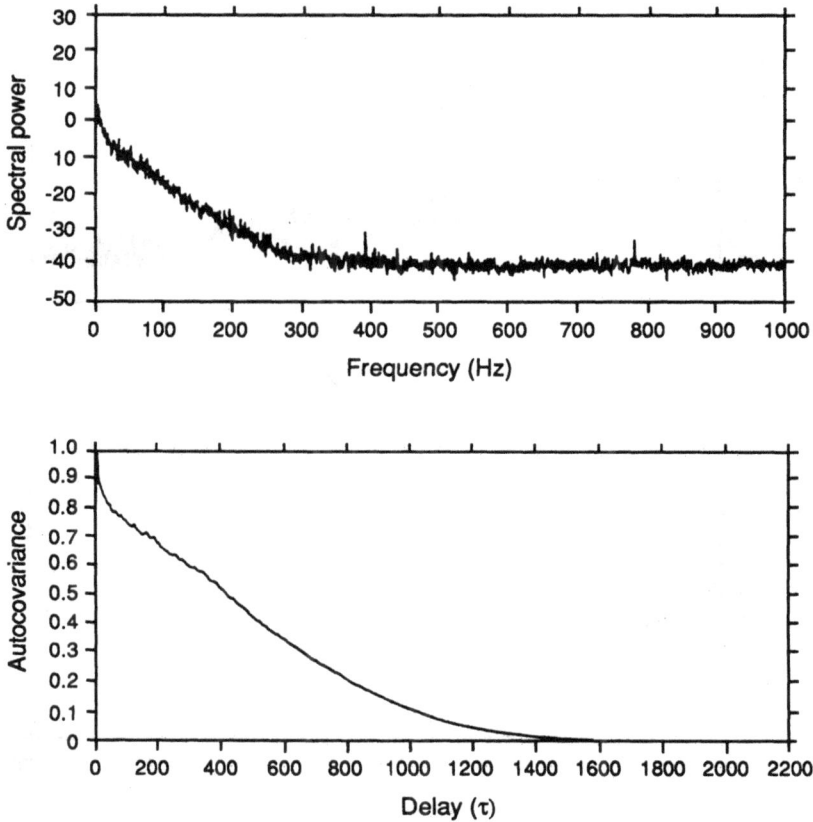

Figure 3.72 Sea return measured spectrum and autocovariance for 17 GHz, hh polarization. (From Trebits and Perry [60].)

spectrum and clutter correlation functions can be determined. The general relationships were discussed in the section describing ground clutter and are listed in Table 3.24

From data presented by Long [4] at X-band we have

$$\frac{A}{B} \text{ ranges from } .3 \text{ to } 10, \text{ average } 1.43 \text{ for horizontal polarization} \quad (3.43)$$

$$\frac{A}{B} = 14, \text{ average for vertical polarization} \quad (3.44)$$

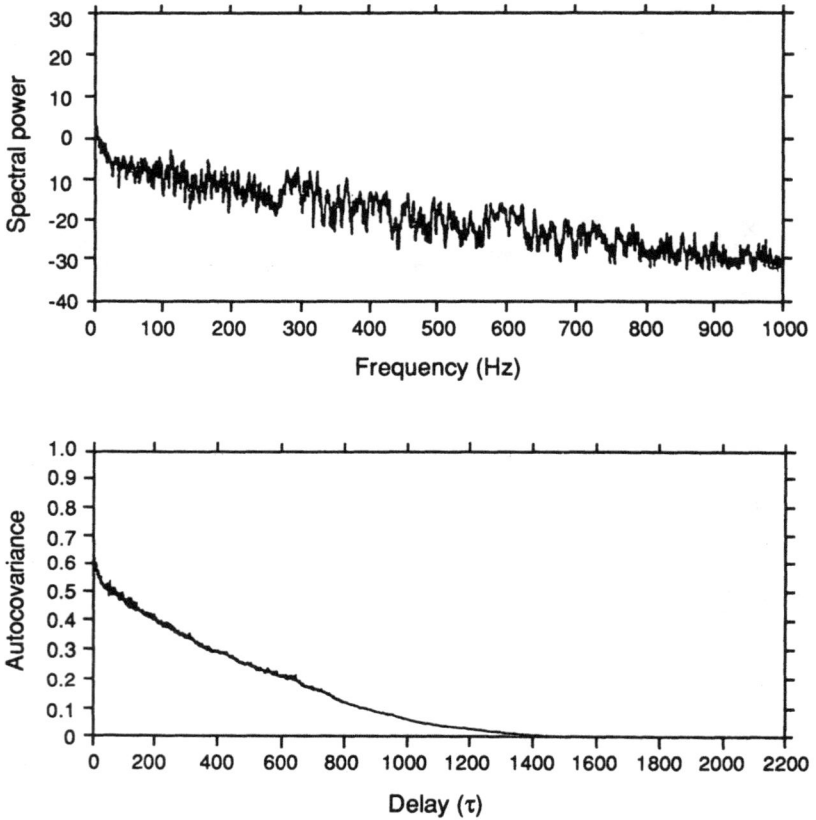

Figure 3.73 Sea return measured spectrum and autocovariance for 35 GHz, hh polarization. (From Trebits and Perry [60].)

$$t_b = 6 \text{ ms, when } e^{-bt} \text{ at half value} \qquad (3.45)$$

$$t_a = 1/2 \text{ sec, when } e^{-\alpha^2 t^2} \text{ at half value} \qquad (3.46)$$

$$t_c = 3 \text{ to 6 sec, period of swells} \qquad (3.47)$$

From the spectrum data presented by Sittrop at X-band,

$$\frac{\pi}{a} = \frac{1}{0.6} \qquad (3.48)$$

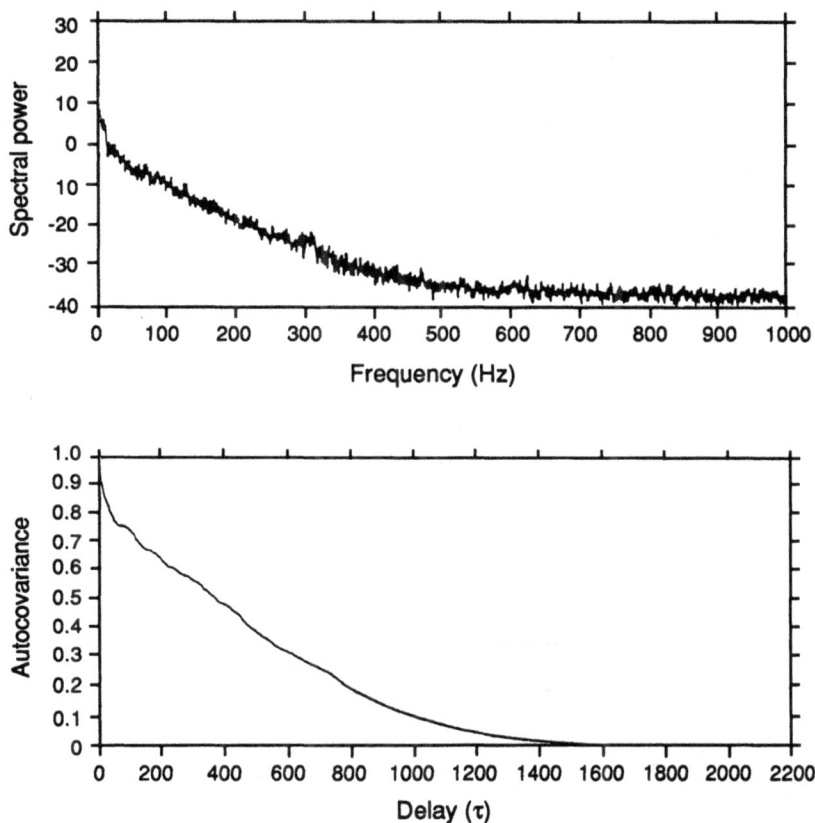

Figure 3.74 Sea return measured spectrum and autocovariance for 95 GHz, hh polarization. (From Trebits and Perry [60].)

Table 3.23
Calculations of the Two Parts of the Autocorrelation Function for Sea Return
(From Trebits and Perry [60])

Autocorrelation Function Component	Radar Frequency			
	95 GHz	*35 GHz*	*17 GHz*	*9.4 GHz*
π/a (1/Hz)	1/100	1/50	1/30	1/10
$2\pi/b$ (1/Hz)	1/.4934	1/.4818	1/.4732	1/.5513

Table 3.24
General Form of the Parts of the Clutter Spectrum for Sea Return

Part	Time	Frequency
a	$Ae^{-\alpha^2 t^2}$	$\left[\dfrac{A2\sqrt{\pi}}{a}\right] e^{-\left(\frac{\pi f}{a}\right)^2}$
b	Be^{-bt}	$\left[B\dfrac{4}{b}\right] \dfrac{1}{1 + \left(\frac{2\pi}{b}f\right)^2}$
c	$C\cos(ct)$	Delta function

Table 3.25
Additional Sources of MMW Clutter Data

No.	Author(s)	Title	Publisher
1	M.W. Long	Backscatter of Land and Sea	Artech House
2	N.C. Currie and C.E. Brown, eds.	Principles & Applications of Millimeter-Wave Radar	Artech House
3	F.T. Ulaby and M.C. Dobson	Handbook of Scattering Statistics for Terrain	Artech House
4	F.T. Ulaby, R.K. Moore, and A.K. Fung	Microwave Remote Sensing Handbook, Vol 3	Artech House
5	J.L. Eaves and E.K. Reedy, eds.	Principles of Modern Radar	Van Nostrand-Reinhold

$$\frac{2\pi}{b} = \frac{1}{80} \tag{3.49}$$

$$\frac{A\dfrac{2\sqrt{\pi}}{a}}{B\dfrac{4}{b}} = 18 \text{ dB} \tag{3.50}$$

3.4 SUMMARY

In this chapter we have summarized the data available in the literature for atmospheric and surface clutter at millimeter waves. The data presented here are not exhaustive since all of the available data would fill many books. If the reader requires additional data, the selected references in Table 3.25 are recommended.

REFERENCES

[1]   A.R. Downs, "A Model for Predicting the Rain Backscatter from a 70 GHz Radar," U.S. Army Ballistic Research Laboratory Report No. 2467, Aberdeen Proving Ground, Maryland, 1975, p. 39.

[2]   D. Turner, "Terrestrial Microwave Radio Relay System Development at Frequencies Above 10 GHz," Private Communications with R.D. Hayes, British Telecommunication Headquarters, Appleton Laboratory, Great Britain, 1975.

[3]   D.B. Ebeoglu, "Millimeter Wave Data Reproducibility, Comparibility and Blending," AC/243 (Panel RSG.8) Land Sub-group Workshop, Eglin AFB, Florida, June 1990.

[4]   M.W Long, *Backscatter of Land and Sea,* Artech House, Inc., Norwood, Massachusetts, 1984, p. 22.

[5]   N.C. Currie, F.B. Dyer, and R.D. Hayes, "Analysis of Radar Rain Return at 9.375, 35, 70, and 95 GHz," Technical Report No. 2 on Contract DAAA 25–76–0256, Georgia Tech, Atlanta, Georgia, February 1975.

[6]   V.W. Richard and J.E. Kammerer, "Rain Backscatter Measurements and Theory at Millimeter Wavelengths," Report No. 1838, U.S. Army Ballistic Research Laboratory, Aberdeen Proving Ground, Maryland, October 1975.

[7]   M.I Skolnik, *Introduction to Radar Systems,* McGraw-Hill Book Company, New York, 1962, p. 33.

[8]   F.E. Nathanson, *Radar Design Principles,* McGraw-Hill Book Company, New York, 1969, p. 200.

[9]   D.K. Barton, *Radar System Analysis,* Artech House, Inc., Norwood, Massachusetts, 1976, p. 105.

[10]   D.W. Beam, Chapter 24 in *Radar Handbook,* M.I. Skolnik, ed., McGraw-Hill Book Company, 1970, p. 655.

[11]   W. Fishbein et al., "Clutter Attenuation Analysis," Technical Report No. ECOM-2808, USAECOM, Fort Monmouth, New Jersey, March 1967, AD665352.

[12]   J.B. Billingsly and J.F. Larrabee, "Measured Spectral Extent of L- and X-Band Radar Reflectivity from Wind Blown Trees, "Lincoln Lab Project Report CMT-57, Massachusetts Institute of Technology, Boston, Massachusetts, February 1987.

[13]   M.W Long, *Backscatter of Land and Sea,* Artech House, Inc., Norwood, Massachusetts, 1984, pp. 192–195.

[14]   N.C. Currie, R.D. Hayes, and F.B. Dyer,"Backscatter From Land Clutter at 10, 16, 35, and 95 GHz," Technical Report No. 3 on Contract DAAA 25–73–0256, Georgia Institute of Technology, Atlanta, Georgia, April, 1975.

[15]   F.E. Nathanson, *Radar Design Principles,* McGraw Hill Book Company, New York, 1972, p. 252.

[16] J.W. Mink, "Rain Attenuation and Side Scatter Measurements of Millimeter Waves Over Short Paths," Technical Report No. ECOM- 4327, U.S. Army Electronics Command, June 1975.

[17] W.P.M.N. Keitzer, J. Sneider, and C.D. de Hann, as adapted by R.N. Trebits, "Millimeter Wave Propagation Phenomena," Chapter 4 of *Principles and Applications of Millimeter Wave Radar,* N.C. Currie and C.E. Brown, eds., Artech House, Inc., Norwood, Massachusetts, 1989, p. 156.

[18] J. Nemarich, R.J. Wellman, and J. Lacombe, "Backscatter and Attenuation by Falling Snow and Rain at 96, 140, and 225 GHz," *IEEE Trans Geoscience and Remote Sensing,* Vol. 26, No. 3, May 1988, pp. 330–342.

[19] V. Furuhama et al., "Experimental Study of Propagation Characteristics in Millimeter Wave Region," *Proc. URSI Commission F1983 Symposium,* June 1983.

[20] R.N. Trebits, "Millimeter Wave Propagation Phenomena," Chapter 4 of *Principles and Applications of Millimeter Wave Radar,* N.C. Currie and C.E. Brown, eds., Artech House, Inc., Norwood, Massachusetts, 1989, p. 152.

[21] V.W. Richard et al., "140 GHz Attenuation and Optical Visibility Measurements of Fog, Rain, and Snow," Report No. ARBRL-MR- 2800, U.S. Army Ballistic Research Laboratory, Aberdeen Proving Ground, Maryland, December 1977.

[22] J. Nemarich et al., "Comparitive Near Millimeter Wave Propagation Properties of Snow and Rain," *Proceedings of Snow Symposium III,* Hanover, New Hampshire, August, 1983.

[23] M.P. Langelben, "The Terminal Velocity of Snow Aggregate," *Quarterly Journal of the Royal Meteorological Society,* Vol. 80, 1954, p. 174.

[24] C.L. Belcher et al., "Millimeter Wave Clutter Reflectivity and Attenuation Handbook," Final Technical Report on Contract DAAH01–84-D-A029–0051, Georgia Institute of Technology, Atlanta, Georgia, March 1987.

[25] E.E. Martin, "Radar Propagation Through Dust Clouds Lofted by High Explosive Tests, MISERS BLUFF Phase II," Final Technical Report on Project A-2465 for SRI International, Georgia Institute of Technology, Atlanta, Georgia, October 1980.

[26] E.E. Martin, "Radar Scattering Properties of the Dust Cloud Lofted by the MISERS BLUFF II High Explosive Test," *Proceedings of the Workshop on Millimeter and Submillimeter Atmospheric Propagation Applicable to Radar and Missile Systems,* Redstone Arsenal, Alabama, March 1979, pp. 109–113.

[27] F.C. Petito, private communication with R.D. Hayes, July 1979.

[28] J.E. Knox, "Millimeter Wave Propagation in Smoke," Conference Record of the IEEE EASCON Meeting, Arlington, Virginia, Vol 2, October 1979, pp. 357–361

[29] R.D. Hayes, "Radar Scattering and Absorption by Sand," unpublished paper, RDH, Inc., Marietta, Georgia, February, 1989

[30] F.B. Dyer and R.D. Hayes, "Computer Modeling of Fire Control Radar Systems," Technical Report No. 1 on Contract DAAA 25–73- 0256, Georgia Institute of Technology, Atlanta, Georgia, July 1974.

[31] T.L. Lane, "95 GHz Clutter Backscatter TABILS V Data Base Summary," Final Technical Report on Boeing Airplane Company Contract No. BB3041, Georgia Institute of Technology, Atlanta, Georgia, December 1982.

[32] F.T. Ulaby and M.C. Dobson, *Handbook of Radar Scattering Statistics for Terrain,* Artech House, Inc., Norwood, Massachusetts, 1989, pp. 257–355.

[33] N.C. Currie, F.B Dyer, and G.W. Ewell, "MMW Radar Reflectivity Measurements From Snow," Technical Report No. AFATL-TR-77–4, Eglin AFB, Florida, April 1977.

[34] W.H. Stiles and F.T. Ulaby, "Microwave Sensing of Snowpacks," NASA Report No. 3263 on Contract NAS5–23777, University of Kansas Center for Research, Lawrence, Kansas, June 1980.

[35]   N.C. Currie et al., "Millimeter-Wave Measurements and Analysis of Snow-Covered Ground," *IEEE Trans. Geoscience and Remote Sensing*, Vol. 26, No. 3, May 1988, pp. 307–318.

[36]   N.C. Currie, "SNOWMAN Radar/Ground Truth Data Analysis," Interim Technical Report on Contract DAAH01–84-D-A029–0014, Georgia Institute of Technology, Atlanta, Georgia, November 1985.

[37]   S.P. Zehner and M.T. Tuley, "Development and Validation of Multipath and Clutter Models for Tac Zinger in Low Altitude Scenarios," Final Technical Report on Contract F49620–78-C-0121, Georgia Institute of Technology, Atlanta, Georgia, 1979.

[38]   N.C. Currie and S.P. Zehner," Millimeter Wave Clutter Model," *Proceedings of the IEE RADAR 82 Conference*, London, September 1982.

[39]   N.C. Currie and S.P. Zehner, "Millimeter Wave Clutter Model Update," *Proceedings of the IEE RADAR 87 Conference*, London, October 1987.

[40]   R.D. Hayes and M.W. Long, "Radar Modeling for MM Wave Applications," Final Report on Contract DAAG29–76-D-0100, Georgia Institute of Technology, Atlanta, Georgia, July 1978.

[41]   E.J. Barlow, "Doppler Radar," *Proc. IRE*, Vol. 37, April 1949, pp. 340–355.

[42]   D.K. Barton, *Radar System Analysis*, Prentice-Hall, 1964, Artech House, Inc., Norwood, Massachusetts, 1976, p. 98.

[43]   M.I. Skolnik, *Introduction to Radar Systems*, McGraw-Hill Book Company, New York, 1962, p.146.

[44]   M.W. Long, *Backscatter of Land and Sea*, Artech House, Inc., Norwood, Massachusetts, 1984, p. 157.

[45]   M.W. Long,"Statistics of RCS Data," Chapter 7 in *Radar Refectivity Measurement: Techniques and Applications*, N.C. Currie, ed., Artech House, Inc., Norwood, Massachusetts, 1989, pp. 222–225.

[46]   Lawson and Uhlenbeck, *Threshold Signals*, Chapter 3, Vol. 24, MIT Radiation Laboratory Series, McGraw-Hill Book Company, New York, 1946.

[47]   W.K. Rivers, Jr., "Detection of Men Carrying Rifles," Final Technical Report on Contract DAAD 05–76-C-0070, Georgia Institute of Technology, Atlanta, Georgia, January 1968.

[48]   N.C. Currie, F.B. Dyer, and R.D. Hayes, "Millimeter Wave Land Clutter Measurements at 10, 16, 35, and 95 GHz," Technical Report No. 3 on Contract DAAA 25–76–0256, Georgia Tech, Atlanta, Georgia, 2 April 1975, ADA 012709.

[49]   R.R. Booth, "The Weibull Distribution Applied to Ground Clutter Backscatter Coefficient," Report No. RE-TR-69–15, U.S. Army Missile Command, Redstone Arsenal, Alabama, June 1969.

[50]   T.L. Lane, "95 GHz Clutter Backscatter TABILS V Data Base Summary," Final Technical Report on Boeing Airplane Company Contract No. BB3041, Georgia Institute of Technology, Atlanta, Georgia, December 1982, p. 42.

[51]   N.C. Currie et al., "Millimeter-Wave Measurements and Analysis of Snow-Covered Ground," *IEEE Trans. Geoscience and Remote Sensing*, Vol. 26, No. 3, May 1988, p. 313.

[52]   W.K. Rivers, "Low Angle Sea Return at 3 mm Wavelength," Final Report on Contract N62269–70-C-0489, Georgia Institute of Technology, Atlanta, Georgia, 1970.

[53]   R.N. Trebits, N.C. Currie, and F.B. Dyer, "Multifrequency Radar Measurements," *URSI Commission F Symposium*, Montreal, Quebec, June 1980.

[54]   M.M. Horst and B. Perry, IV, "MMW Modeling Techniques," Chapter 8 in *Principles and Applications of Millimeter-Wave Radar*, N.C. Currie and C.E. Brown, eds., Artech House, Inc., Norwood, Massachusetts, 1987.

[55]   M.M. Horst et al., "Radar Sea Clutter Model," *IEEE APS Digest*, Part 2, 1978.

[56]   G.W. Ewell, M.M. Horst, and M.T. Tuley, "Predicting the Performance of Low Angle Mi-

crowave Search Radars—Targets, Sea Clutter, and the Detection Process," *Proceedings of OCEANS,* September 1978, pp. 373–378.

[57] M.M. Horst, F.B. Dyer, and M.T. Tuley, "Radar Sea Clutter Model," *URSI Digest,* 1978 IEEE APS/URSI International Symposium, College Park, Maryland, 1978.

[58] R.D. Hayes et al., "Some Statistical Properties of Radar Returns From the Sea," Georgia Tech Project A-747, Naval Air Development Center, 1965.

[59] H. Sittrop, "X- and K_u-Band Radar Backscatter Characteristics of Sea Clutter," Parts I and II, Physics Laboratory of the National Defense Research Organization, The Hague, The Netherlands, 1975.

[60] R.N. Trebits and B. Perry, "Multifrequency Radar Sea Backscatter Data Reduction," Georgia Tech Project A-2717, Naval Surface Weapons Center, 1982.

Chapter 4
Detection of Targets Immersed in Clutter

4.1 INTRODUCTION

In this chapter we will summarize the process used by a millimeter-wave system designer in developing the characteristics of a system to solve a specific problem and then in analyzing the performance of the postulated system. In no way will this discussion be exhaustive, as literally thousands of books, papers, and reports have been written on the subject. Rather, the goal is to provide the reader a glimpse of the thought processes involved in such a design along with supplying a list of references for further reading.

The development of a system begins with the definition of the key requirements and constraints that relate to the postulated system. These requirements and constraints are summarized in Table 4.1. From the requirements, such as the mission, scenario, target, geometry, and environment, secondary parameters for the system are developed, such as target RCS, clutter type and reflectivity coefficient (σ^0), atmospheric attenuation (α), and required system resolution. From constraints imposed by either physical laws or the state of the art for radar components, limitations on search windows, dwell times, minimum detectable signal, etc., determine how well the radar can meet the design requirements. One of the variables under the designer's control is the frequency. In particular, for small, fixed aperture systems such as those located on a missile or airframe, the optimum choice is often millimeter wave, even though environmental effects may be more severe than at lower frequencies.

The basic procedure that a radar designer could typically follow in developing a new system is illustrated in Figure 4.1. At the start, decisions are made as to transmitted wavelength, aperture diameter, gimbal limits, volume, and other constraints on the system/antenna. Based on these preliminary assumptions, a calculation of the expected position accuracy is made. If this accuracy is not accomplished, the initial assumptions are modified (usually raising the frequency) until the accuracy requirement is satisfied. Then the pulsewidth (or equivalent frequency-modulated constant wave (FMCW) bandwidth) is determined based on the maximum

Table 4.1
Requirements and Constraints for a Radar Design

Requirements	Constraints

1. Mission
  a. Mapping
  b. Fire control
  c. Seeker
  d. Instrumentation
2. Scenario
  a. Air-to-ground
  b. Air-to-air
  c. Ground-to-air
  d. Ground-to-ground
3. Target
  a. Aircraft
  b. Missile
  c. Vehicle
  d. Boat/ship
4. Geometry
  a. Radar altitude
  b. Target altitude
  c. Minimum and maximum required
     detection range
  d. Search window
  e. Radar platform speed
  f. Target speed
5. Environment
  a. Desert
  b. Temperate
  c. Tropical
  d. Arctic
  e. Sea
	1. Maximum aperture size
	2. Maximum size (volume) and weight for the
	   system
	3. Maximum peak and average transmitted power
	4. Cost and reliability (related to complexity)
	5. Maximum antenna scan rate (versus required
	   search volume)
	6. Scan limits
	7. Measures of Performance (P_D, P_{FA}, etc.)

target dimension. (Generally, the maximum signal-to-clutter ratio is achieved when the pulse length matches the target length.) Based on the assumed pulse length, the receiver bandwidth is calculated. Next, the pulse repetition frequency (PRF) is determined to provide unambiguous range or Doppler spectrum based on the range and velocity of the target. Note that a key tradeoff for a radar using moving target indicator (MTI) is unambiguous range versus unambiguous target Doppler. Sometimes a compromise is used in which both are ambiguous. Stimson [2] and Morris [3] contain excellent discussions of this tradeoff. Next, the scan rate is determined from the requirements for scan volume, scan time, and search pattern, and, finally, the

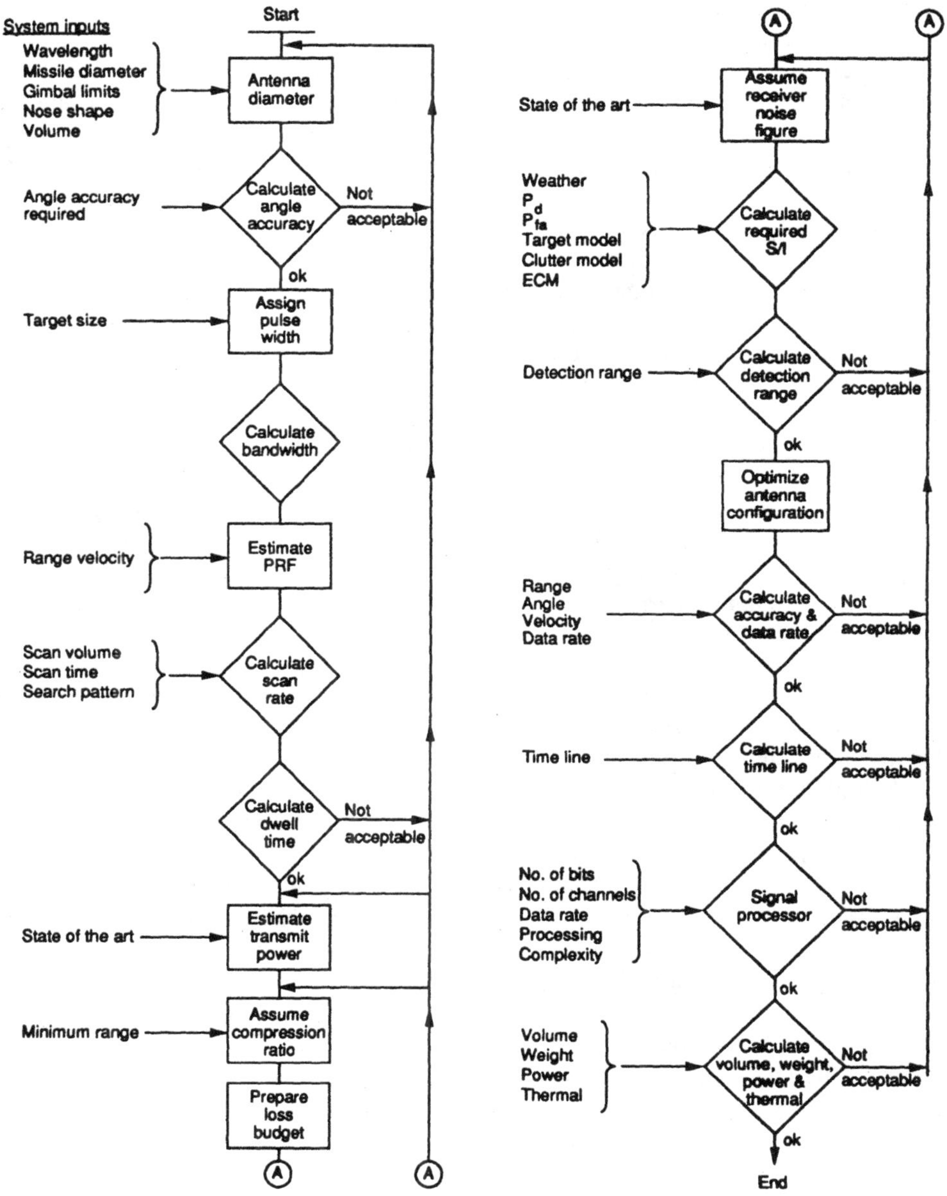

Figure 4.1 Idealized flow chart for radar system design methodology. (From Scheer [1].)

dwell time on a specific radar cell is calculated. From the dwell time and PRF, the number of pulses that can be integrated to improve detection is calculated.

Continuing the design, the state-of-the-art limits for transmit power for a given frequency and size are determined, and any gain in effective transmit power due to pulse compression, taking into account that the required minimum range for the system is maintained. Next, the system losses (known as the loss budget) are determined, including component, waveguide, antenna, and radome losses. A noise figure for the receiver is determined from receiver components, and, based on the requirements for probability of detection, probability of false alarm, target models, and clutter models, a required signal-to-noise ratio (SNR) is determined. Using the radar range equation, the maximum detection range for the target is calculated. If the range is not acceptable, the previous parameters are modified until the required range is obtained. When this occurs, the antenna configuration is optimized, accuracy and data rates are determined based on typical equipment specifications, the signal processor loading is estimated, and finally the required volume, weight, power, and thermal dissipation limits are estimated to ensure that none of the constraints were exceeded. If any constraints were exceeded, the process iterates until all are satisfied.

The key analysis of maximum range for a given set of radar parameters uses the radar range equation that was introduced in Chapter 1. This equation describes the basic performance of a given radar. In analyzing the radar equation, there are three basic groups of parameters that affect the received energy from a reflecting surface. The system parameters include radiated power and wave form management, frequency, bandwidth, antenna gain, polarization, and signal processing, which have a direct bearing on the ability for the system to differentiate the desired signal from noise and clutter. The second group relates to the path between the transmitter and the receiver. In this group consideration is given to signal losses (absorption and scattering) due to rain, snow, dust, molecular gases, multiscattering surfaces, and range between the transmitter and the receiver. The third group is concerned with the reflecting surfaces or volumes, usually referred to as σ. Whether the reflecting object is a point object, area extended, or a volume is relative to the cell of resolution of the radar. The cell in turn is controlled by system parameters such as antenna beamwidths and transmitter pulse length or modulation waveform.

There are physical and molecular parameters used to describe each of the parameters in each of the three groups. For example, reflecting objects scatter the arriving signal depending on object dimensions, number of discrete parts, curvature, surface roughness, dielectric constant, permeability, emissivity, transparency, density, temperature, pressure, polarization, motion, and radar frequency. These physical and molecular parameters cause the radar received signal scattered by the object to change the absolute magnitude of the scattered signal, to have amplitude distributions with time and space (radar resolution cell), to vary in time with both a modulation spectrum and a group frequency, to have phase coherent or noncoherent response, to be sensitive to signal polarization, and to permit a portion of the signal to pass through the object (rain, snow, dust, and fog). With the relationships of the physical and molecular parameters to the scattering object (the third group), the radar engineer can design the system parameters (the first group) to maximize the detection of an object against noise, identify the desired object within a large group of detected objects, and classify the object as a specific object of interest in all of the objects detected in the field of view.

There are four domains in which most radar systems are designed to take advantage of scattering object characteristics in order to distinguish desired objects from interference and clutter. These domains are amplitude, time, frequency, and polarization.

This chapter will present some examples of the detection process and the impact on system parameter design.

4.2 OVERVIEW OF THE DETECTION PROCESS*

This chapter presents an overview of the radar system design considerations allowing the radar designer to make the decisions associated with selection of the various radar system design parameters. The material presented here demonstrates how the attenuation and reflectivity data presented in Chapters 2 and 3 may be applied to a radar design.

Although a type of design methodology is presented here, it must be realized that there is no "cookbook" approach to the millimeter-wave radar design process. Several inter-relationships do exist in developing the system parameters, such as the relationship among PRF, unambiguous Doppler, and unambiguous range.

The radar range equation is presented again, this time from a designer's point of view. Antenna, transmitter, receiver, and processing topics are covered as they relate to the development of the specified performance of these functions. Various signal processing concepts and their relationship to the overall system performance are also presented.

The sequence of thought processes through which a radar system designer might progress was presented in Figure 4.1. The most important initial decision for the radar, the waveform desired, is often relatively easy to select. Table 4.2 gives a list of several salient radar system performance specifications and lists the key radar parameter(s) driving each particular performance parameter. The discussions to follow will provide the reader with an understanding of the detailed relationships between the desired parameters and the required radar specifications.

---

*Authors note: Portions of this section were adapted with permission from Scheer [4], pp. 616–625.

Table 4.2
Key Radar System Relationships

Given Specification	Radar Parameter	Implied Performance
Range resolution	Pulsewidth	Bandwidth
	System Bandwidth	Signal-to-interference performance
Range (max)	PRF	Integration gain
		Unambiguous Doppler
Target velocity (Doppler detection)	PRF	Integration gain
		Unambiguous Doppler
Cross-range resolution	Beamwidth	Antenna gain
Detection probability	SNR	Range of detection
Accuracy	SIR	Resolution, processing
Target evaluation	Processing	Resolution coherence

4.2.1 Amplitude Detection

Of the domains used by radar engineers to detect and identify targets, the amplitude domain was the first and is still very important to distinguish the desired signal from noise and some other interference. When there are several signals originating from different sources, the detection of the desired signal in the packet of signal plus interference will result in some statistical analysis. The usual approach in radar system signal detection is to determine the probability of detection, with a probability of false alarm, when signal, clutter, and noise are all in a particular cell of resolution. Figure 4.2 depicts three such signals. A circle has been drawn to represent the average value, and the larger the circle, the larger the average value. Variations about the circle represent the fluctuations described by some density function. The receiver signal is presented in the complex plane to show in-phase (I) and quadrature-phase (Q) and, thus, the phase angle of the signals.

The resulting signal at any instant of time is the vector addition of the three signals. In the pure sense of signal detection, the detected signal will appear at any position in the complex plane. The power value is the absolute value of the resultant signal squared.

For the case of a large SNR, the diameter of the signal is large and the diameter of the noise is small. The variation of the received signal is representative of the desired signal with a small fluctuation due to the relatively small variation caused by noise. The rate of rotation of the total signal will represent a moving target, and this rate of rotation (or phase shift) is the Doppler frequency.

Chi-square target

K-distributed clutter

Noise

Complex single return in cell containing target, clutter, noise

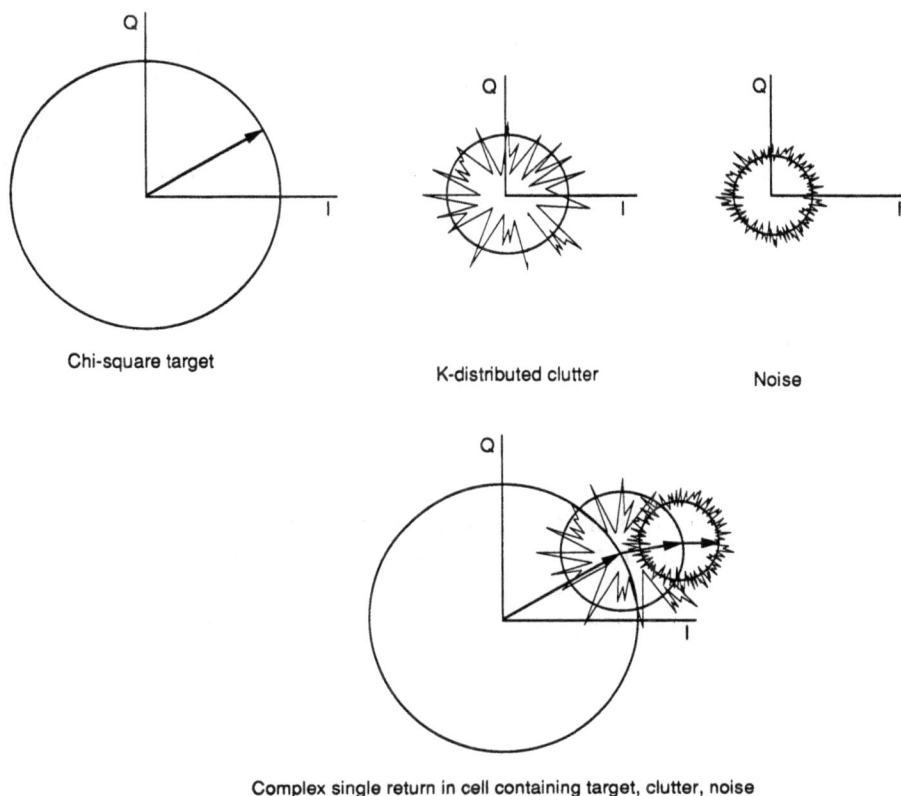

Figure 4.2 Representation of the phasor vectors and the resultant for target signal, clutter, and noise.

It is known from theory presented in Chapter 2 and data presented in Chapter 3 that millimeter waves are limited to shorter ranges (less than 10 km) because of higher atmospheric losses. It is also well known from antenna theory that the higher the frequency (the shorter the wavelength) the smaller the beamwidth will be for a given antenna aperture, and the smaller the beamwidth, the smaller the intercepted area and the volume of extended targets. These smaller beams will produce less interference from unwanted scatterers, such as ground, sea, and rain clutter. The application of millimeter-wave radar systems, then, is to employ some guidance technique to bring the radar to a point in space any place on the earth where terminal operations of fine resolution in an adverse environment are required.

The final operation may be to land an aircraft at an airport in fog where there is zero ceiling and zero visibility, or to deliver a missile on an object in a crude battle environment. The consideration here is the impact of ground clutter on the detection process of a millimeter radar to detect a desired surface or object in an area where clutter will cause false targets and the millimeter radar to miss designating the desired target.

Coast and geodetic survey maps and aerial maps of central Europe have been reviewed in mixed urban and rural areas, not deserts and not vast farm lands in mid-America. It is observed that when considering a strip 6000 m down range by 1000 m cross range, the scene will typically be composed of

30%	woods
45%	cultivated land and fields
15%	man made structures
10%	ponds, lakes, and roads

In review of data presented in Chapter 3, the radar cross section (RCS) per unit area at 95 GHz and 30° grazing angle will have typical values of

$\sigma^0 = -25$ dB (m^2/m^2)	Ponds, lakes, roads
$\sigma^0 = -20$ to -12 dB	Soil and seedlings
$\sigma^0 = -13$ to -9 dB	Crops
$\sigma^0 = -10$ to -5 dB	Trees
$\sigma^0 = -5$ to $+5$ dB	Urban

When surveillance is at a grazing angle of less than 5°, then the vegetation-type surfaces (soil, grass, crops, tree foliage) will have backscatter coefficients some 20 dB less than the values listed above at a 30° depression angle. Tree trunks and urban areas are not as sensitive to grazing angles, and will be in the range of -5 to $+5$ dB (m^2/m^2).

The distribution of the ground vegetation radar backscatter is bimodal and will typically be found to appear as shown in Figures 4.3 and 4.4. When the lakes and the urban areas are added, then the distribution of cross sections will be trimodal. Such multimodal distributions have been modeled as Weibull and K functions. Beebe and Yoshitani have found the amplitude of like values to be clustered in space, and the spatial distribution in terms of nearest neighbor's distance of false targets (not desired target) to be a Poisson distribution [5].

At the 30° depression angle, the wide variation in backscatter coefficients of -25 to $+5$ dB (m^2/m^2) is approaching the dynamic range of present-day linear receivers. When large manmade targets are also in the scene, some technique, such

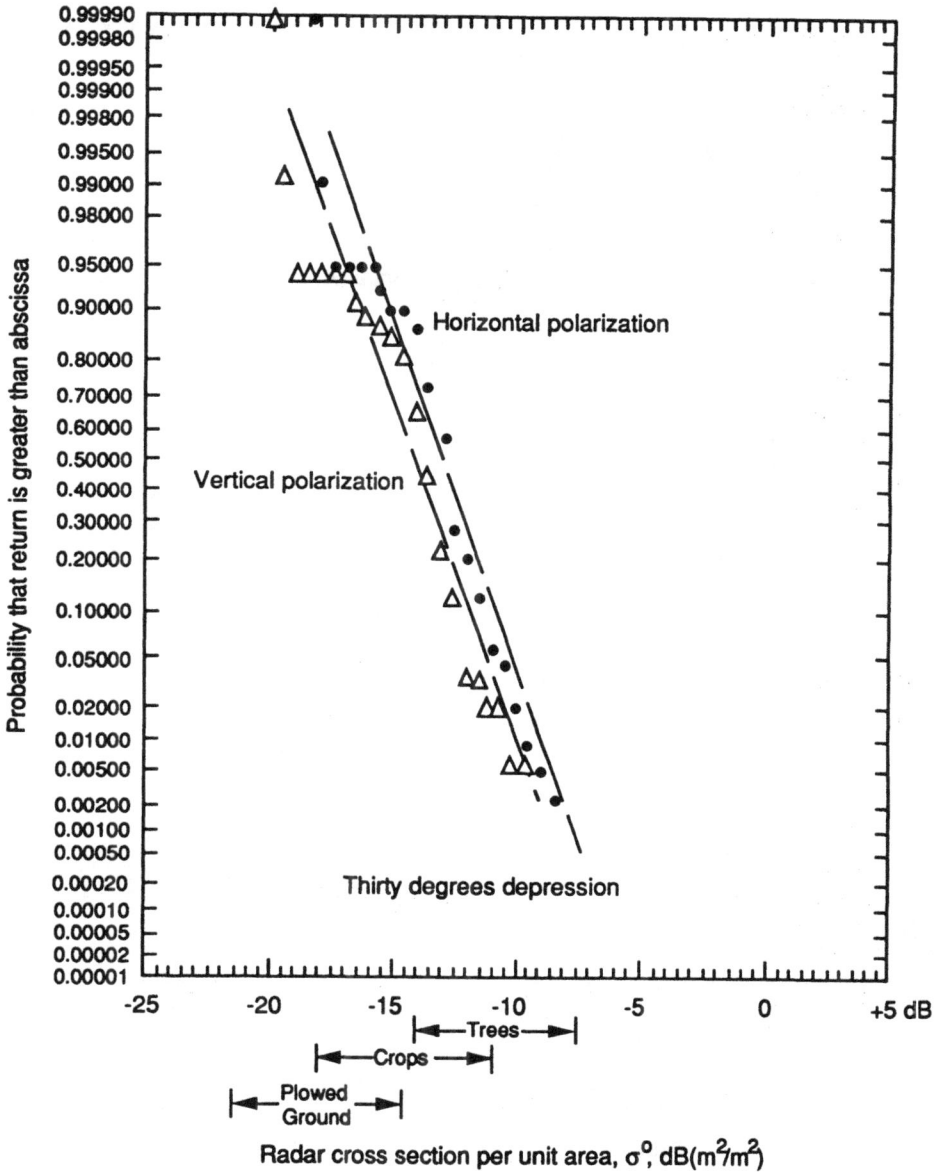

Figure 4.3 Distribution of ground clutter of various types at 30° depression angle for 35 GHz.

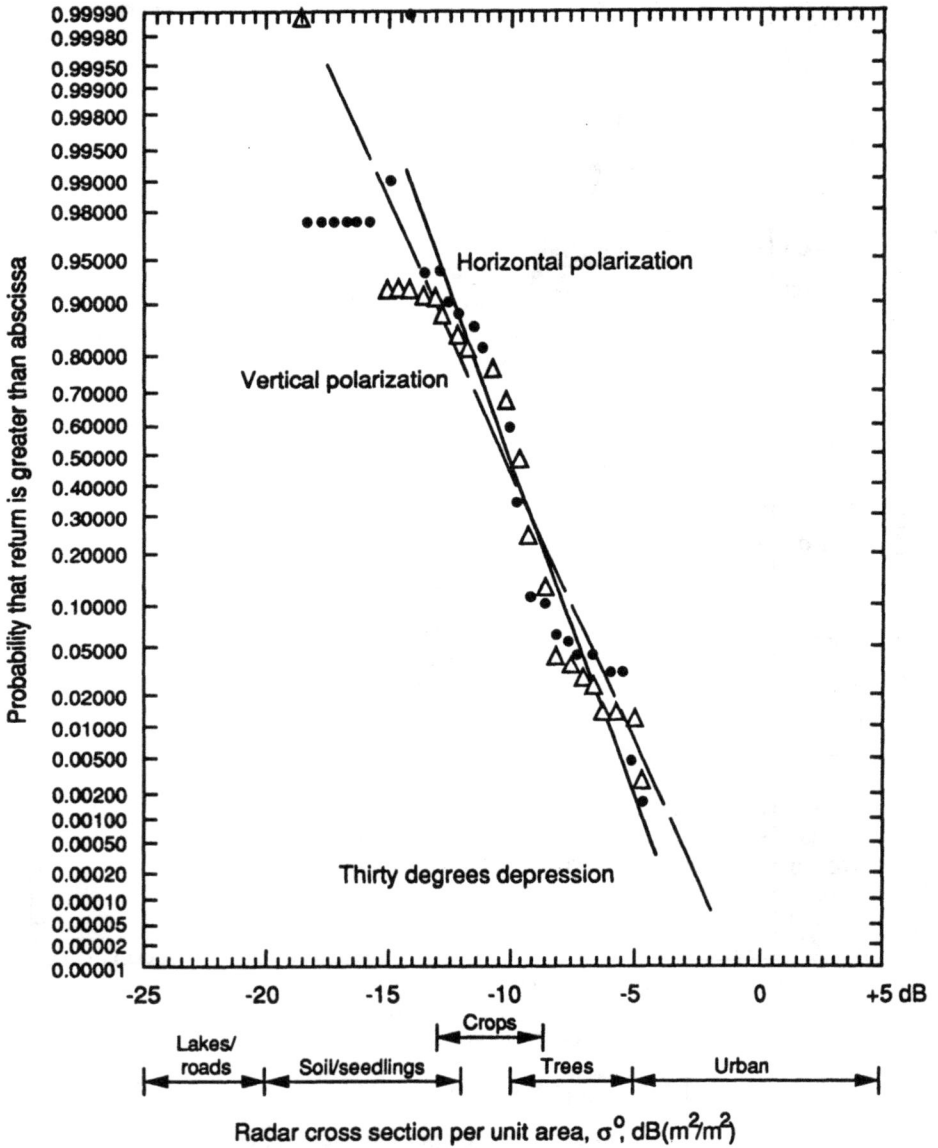

Figure 4.4 Distribution of ground clutter of various types at 30° depression angle for 95 GHz.

as automatic gain control, is required to extend the dynamic range of the linear receiver. And when surveillance is over a larger differential range, then some technique, such as sensitivity time control or a cosecant squared shaped antenna beam, is required to compensate for the signal variation due to range.

The most basic measure of the performance of a radar system is its signal-to-interference ratio (SIR). Chapter 3 presents an analysis of the detection of targets in interference. The discussion here addresses the analysis of target detection in the presence of clutter and noise from a more practical point of view, primarily for the benefit of the system designer.

4.2.1.1 Interference Due to Noise

The single-pulse signal power returned from a target may be accurately predicted from the radar range equation

$$P_r = [P_t G^2 \lambda^2 \sigma]/[(4\pi)^3 L_s L_{atm} R^4] \tag{4.1}$$

where

P_r = received power
P_t = peak transmitted power
G = antenna gain
λ = wavelength
σ = target cross section
L_{atm} = the atmospheric loss as a function of range R (two-way)
L_s = system loss (>1.0)
R = range to the target

The noise power P_n with which the target signal must compete is also predictable from

$$P_n = kTF_n B_n \tag{4.2}$$

where

k = Boltzmann's constant (1.38×10^{-23} J/K)
T = receiver temperature (290 K)
F_n = receiver noise figure
B_n = receiver noise bandwidth

The resulting SNR, assuming the entire received spectrum falls within the receiver bandwidth, is found from

$$\text{SNR} = [P_t G^2 \lambda^2 \sigma] / [(4\pi)^3 kTF_n B_n L_s L_{\text{atm}} R^4] \tag{4.3}$$

Figure 4.5 is a curve showing the SNR as a function of range for a set of typical parameters for 35- and 95-GHz radar systems having the same signal-to-noise performance at 2.5 km. Notice that the slope is −12 dB per decade, plus the effects of atmospheric attenuation. The −12-dB part of the rolloff is predictable from the R^4 term in the denominator of the radar equation. The severe effect of the atmospheric attenuation on the 95-GHz system relative to that for the 35-GHz system is evident from Figure 4.5. Table 4.3 gives the parameters for the two postulated millimeter-wave systems.

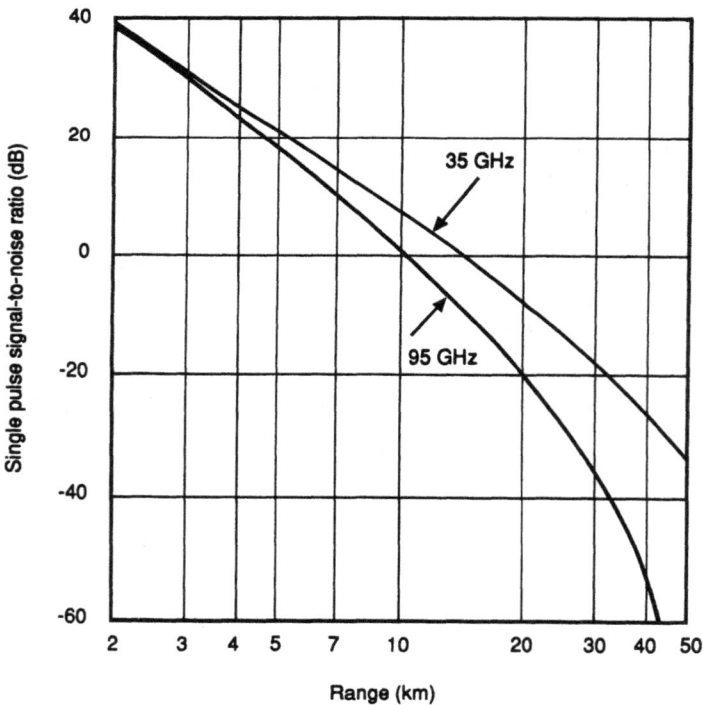

Figure 4.5 Single-pulse SNR for 35- and 95-GHz systems, with parameters given in Table 4.3.

Table 4.3
Radar Parameters for SNR Calculation

Parameter	35-GHz Value	95-GHz Value
P_t	1000W	1000W
G	38 dB	43 dB
λ	8.57 mm	3.16 mm
σ	10 m^2	10 m^2
B_n	1 MHz	1 MHz
F_n	8 dB	8 dB
L_s	7 dB	7 dB
L_a	0.3 dB/km	1 dB/km

4.2.1.2 Signal-to-Noise Requirements

A system designer needs to know the relationship between the SNR and the performance of a radar system. There are some relatively sophisticated techniques for determining these relationships, some of which have been reduced to plots that can be used to relate the SNR to the probability of detection and of false alarm. Although the variety of target and interference conditions for which the plots are developed is limited, these plots provide a very good first-level approximation of system performance [6, 7].

A system designer can determine the expected SNR of his system using the radar range equation and the generally accepted rules of thumb for approximating the various terms. A key remaining question concerns the SNR required for a given system application. The basic system performance parameters affected by the SNR are the probability of false alarm (P_{FA}) and the probability of detection (P_D). The requirements for these parameters are normally determined from mission analysis. A system is normally set up for a given false alarm rate, often established using a constant false alarm rate (CFAR) function. Once the false alarm rate is established, the P_D is a direct function of the SNR; that is, as the SNR increases, so does the P_D. The relationships between P_D and SNR are somewhat complex, depending on the probability density functions and the fluctuation statistics of the target and interference signals. Fortunately, there exist several references that plot a variety of conditions representing typical applications. Blake shows curves for several conditions that may represent a designer's scenario [6]. A more complete set of curves has been generated by Meyer and Mayer [7]. Using these curves, a designer can determine his SNR requirements for a wide range of false alarm numbers (defined as the number of detection opportunities in the interval that gives a P_{FA} of 0.5) and target fluctuation conditions, with and without noncoherent integration processing. Improvement based on coherent integration gain is not plotted because it is simple to calculate it manually.

Figure 4.6 illustrates the detection process for a CFAR processor. The solid curves represent the probability density distributions for noise alone and the target plus noise. The desire is to place an amplitude threshold V_T along the x-axis such that the desired P_{FA} (area under the noise curve) is obtained for the clutter alone while maximizing the P_D (area under the target-plus-noise curve). Integrating N samples of the noise and target-plus-noise curves has the effect of narrowing the variance of the distributions, causing less overlap between the curves and making it possible to set a threshold with better detection performance for a given P_{FA}. In the limit, if an infinite number of pulses could be integrated, the noise and target plus noise distributions would become single valued at their average value, and thus no overlap would exist, making it possible to achieve a P_D of 1 with a P_{FA} of 0.

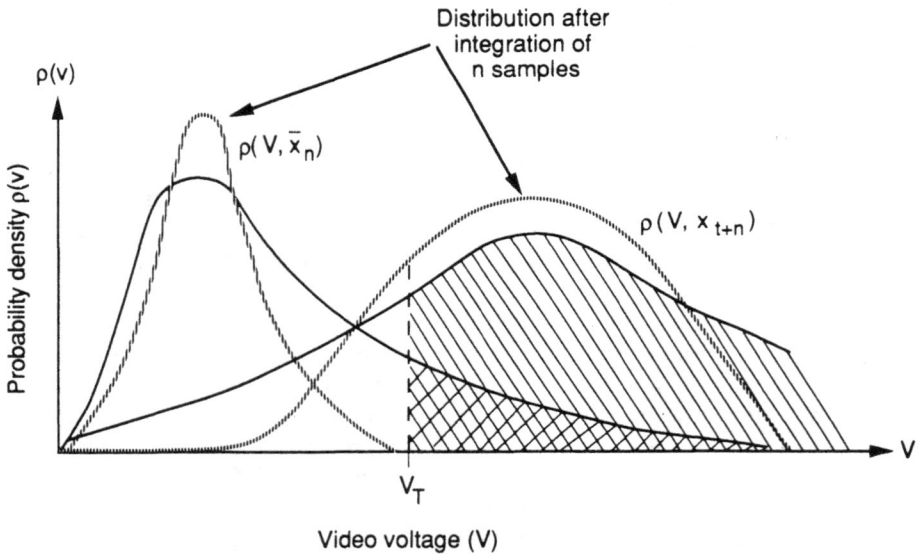

Figure 4.6 Relationship of the probability of detection (PD) to the probability of false alarm (PFA) for single pulse and integration of n pulses cases.

Meyer and Mayer have generated a number of curves that relate the P_D to the false alarm number for a given Swerling target model. The reader should be cautioned that the Meyer and Mayer false-alarm number is not the reciprocal of the false-alarm probability, but is rather the number of detection opportunities in the interval which gives $P_{FA} = 0.5$. The Swerling target model provides for four cases of target fluctuation, which, along with the nonfluctuating case (sometimes called Swerling 0 model), has been accepted by the radar community as descriptive of most types of target fluctuations. The Meyer and Mayer curves are used in several ways, including determination of the required output SNR for a given P_D and P_{FA}. The curves can also be used to determine the input SNR required for a desired P_D if integration is employed.

Note that noncoherent integration has reduced efficiency when compared to coherent integration. However, the higher the input SNR is, the higher is the efficiency of noncoherent integration, which approaches that of coherent for very high SNR. Nonfluctuating targets require the lowest SNR for a given P_D, P_{FA} performance. For fluctuating targets, the required SNR approaches 20 dB for typical cases requiring 90% P_D. For some targets exhibiting non-Rayleigh fluctuation statistics, the Swerling models do not accurately predict SNR for a given P_D and P_{FA}. For these situations, lognormal or Weibull target model fitted to actual data must be used to calculate P_D and P_{FA} as a function of SNR.

4.2.1.3 Interference Due to Clutter

Calculating Clutter Effects

Two forms of natural interference restrict the performance of a radar system designed to detect targets. The most obvious form is the receiver noise, which is highly predictable, as discussed above; the other is clutter backscatter. Terrestrial (ground), sea, and atmospheric clutter have a strong influence on the performance of millimeter-wave radar systems, and, as such, have a strong influence on the design parameters.

Received clutter power is a function of range in much the same fashion as target power. The RCS of clutter is a function of how much clutter is illuminated and received in a given range gate. Although ground clutter is not usually a homogeneous area scatterer, for purposes of first-order analysis, it is often treated as such, the cross section depending on the area defined by the beam shape and pulsewidth. This treatment of ground clutter characterizes it in terms of average RCS per unit of illuminated area. At millimeter-wave frequencies, typical backscatter RCS values for ground clutter range from -10 to -30 dB (m^2/m^2), with extremes ranging from $+5$ to -50 dB (m^2/m^2) for special cases.

The RCS, σ_c, from an area clutter cell as discussed in Chapter 1 is

$$\sigma_c = A\sigma^0 \qquad (4.4)$$

where

A = illuminated area
σ^0 = clutter backscatter coefficient

The illuminated area is found by defining the range extent of the clutter cell (based on either the pulse length or the beam shape) and cross range extent based on the antenna azimuth beamwidth. For ground (or sea) clutter, the area illuminated is either nearly an ellipse, defined by the azimuth and elevation beamwidths, or a rectangle (or a close approximation) defined by the azimuth beamwidth and the pulsewidth.

In either case, the cross-range dimension, D_c, for small beamwidths is approximated by

$$D_c = \frac{R\theta_{az}}{\alpha \cos \phi} \tag{4.5}$$

where

θ_{az} = antenna beamwidth (radians)
R = slant range
α = beamshape loss = 1.33 (Gaussian beamshape)
ϕ = grazing angle

providing expressions for the effective beam area, A, of

$$A = \frac{R\theta_{az} c\tau/2}{\alpha \cos \phi} \tag{4.6}$$

for the pulse-limited case and

$$A = \frac{\pi R^2 \theta_{az} \theta_{el}}{4\alpha^2 \cos \phi} \tag{4.7}$$

When substituting the expression for clutter reflectivity in place of the target cross section in the radar equation, the nature of the relationship between range and SNR can be seen to change. In particular, as contrasted with $1/R^4$ variation for received target power, area clutter received power varies approximately as $1/R^3$ (except for the usual decrease in σ^0 with decreased depression angle) for the pulse-limited case, and approximately as $1/R^2$ for the beam-limited case. This has major significance for millimeter-wave radars operating in a look-down mode when searching for stationary targets in ground clutter, since actual performance in clutter will always be poorer than predicted using noise statistics.

Atmospheric clutter, such as rain, is a volumetric scatterer, depending on the volume defined by the azimuth and elevation beamwidths and the pulsewidth. The volume defined by these parameters as discussed in Chapter 1 is approximated by

$$V = \frac{\pi R^2 \theta_{az} \theta_{el} c\tau}{16} \tag{4.8}$$

When examining the range dependence of received power from volumetric clutter, it is found to vary as $1/R^2$.

The reduced range dependence for area and volumetric clutter means that the signal-to-clutter ratio decreases with increasing range. For this reason, in many cases, the radar slant range performance, in terms of detecting a 1-m^2 target, becomes limited by clutter interference long before it becomes limited by thermal noise.

Calculating Antenna Parameters

Two antenna parameters are of primary importance to the radar designer: beamwidth and gain. Beamwidth determines the radar resolution and the illuminated clutter, while gain is squared in the radar equation, making it the single most important sensitivity parameter. Simple equations for estimating beamwidth and gain are given below.

The radian beamwidth (BW) of a pencil beam antenna can be estimated from

$$BW(rad) = 1.31\lambda/D \tag{4.9}$$

where

λ is the wavelength
D is the antenna diameter in the same units

The beamwidth in degrees is estimated using

$$BW(deg) = 75\lambda/D \tag{4.10}$$

For most applications and antenna designs, an estimated antenna efficiency factor of 67% will suffice to describe the antenna performance. Based on an efficiency factor of 67% and 27-dB sidelobes, the gain of an antenna may be estimated from

$$G = 28{,}000/(\theta_{az}\theta_{el}) \tag{4.11}$$

where

θ_{az} = azimuth beamwidth, in degrees
θ_{el} = elevation beamwidth, in degrees

For a radar that scans in azimuth, the number of pulses that can be integrated to improve the SNR, or can be processed to determine Doppler characteristics, is limited by the number of pulses transmitted while the antenna is pointed at the target of interest. The dwell time, T_d, is found from

$$T_d = \theta_{az}/\Omega \tag{4.12}$$

where Ω is the scanning rate, and the resulting number of related pulses, n, is

$$n = T_d \text{PRF} \tag{4.13}$$

Since the antenna gain normally specified is that at the center of the beam, and some of the pulses integrated occur at times when the effective gain is less than the peak, there is a scanning loss factor or beam shape loss, on the order of 1.45 (1.6 dB), which reduces the overall integration gain of the system [8].

Detection in Non-Rayleigh Clutter

One convention says that a signal is detected in a noise background when the signal exceeds the peak noise at least 90% of the time. This is an SNR of 8 dB. The technique for making this measurement is known as *tangential sensitivity,* and is depicted in Figure 4.7.

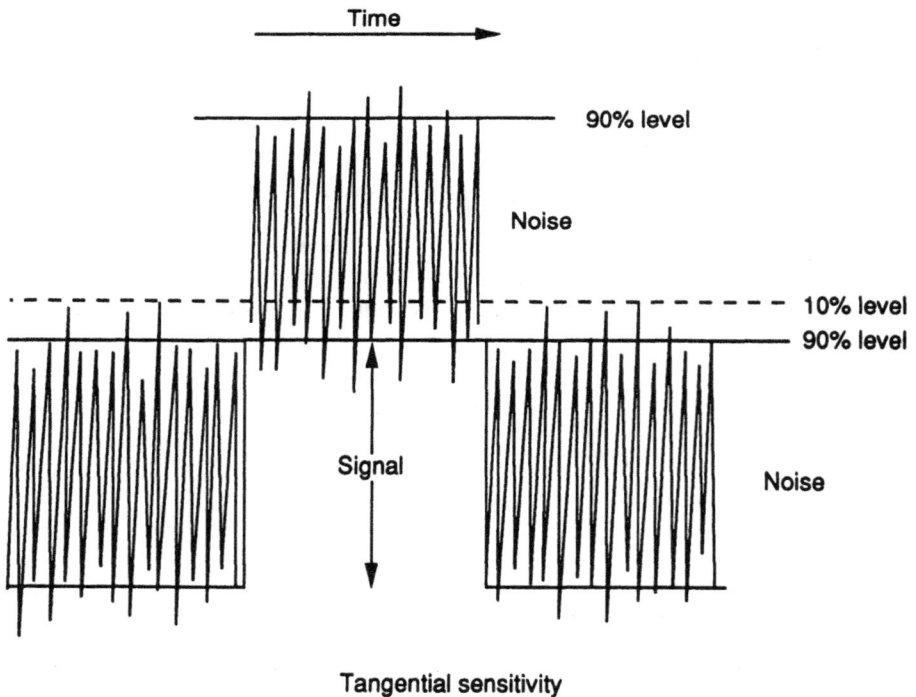

Figure 4.7 Illustration of the definition of tangential sensitivity.

The noise amplitude is assumed to be Gaussian distributed, with an average value as given in Equation (4.2). A rectangular pulse is introduced in the noise signal such that the signal is of a magnitude equal to 90% of the peak noise. The noise will ride on top of the signal, with 10% of the peak-to-peak values extending below the signal itself. From tables of the Gaussian (normal) distribution, it is observed that 90% of the area of the Gaussian distribution is well within the area represented by values within 2.5 standard deviations, or 2.5 times the rms value, from the average. The ratio of signal power-to-noise power is then given by 2.5 squared, which equals 6.35, or 8 dB.

If the signal is not a rectangular signal, but varies in amplitude as a Rayleigh distribution, then the Rayleigh exceeds its average value 37% of the time and exceeds 8 dB below the average value only 85% of the time. In order to have a 90% presence of the Rayleigh signal, the ratio needs to be 10 dB. In Chapter 3, it was discussed that many times the clutter (sea, trees, ground covered snow) has amplitude distributions that are not Rayleigh, but exhibit more complex distributions at millimeter wavelengths. Referring to Figure 4.8 for other distributions, the following values are observed for signals with the same average values.

Distribution	% above average	S/N for 90%
Gaussian	50	8.0 dB
Rayleigh, $b = 1.0$	37	10.0 dB
Weibull, $b = 0.5$	23	22.5 dB
Lognormal, SD = 5 dB	27	10.0 dB
Lognormal, SD = 8 dB	20	18.0 dB

Figure 4.8 also shows that many false alarms will occur if a threshold is set to eliminate the Rayleigh distributed-type scatter most of the time. Consider that a threshold has been set at 9.5 dB above the average return of the Rayleigh scatterers (10^{-4} P_{FA}). The Weibull and lognormal type scatterers will exhibit P_{FA}'s of 5×10^{-3} to 2×10^{-2}, resulting in higher false alarm rates than desired.

As noted by Rivers [9], for the Weibull distribution (almost independent of the form factor b, which includes the Rayleigh) and the lognormal distribution (almost independent of the standard deviation), the average values are 3.5 ± 0.5 dB below the strongest 10% of the magnitudes of the scattering surfaces.

We have established that "complex" distributions such as lognormal and Weibull (two parameter variables) require a higher SNR to have the same P_D, compared to the Rayleigh distribution (one parameter variable). Detection performance can be estimated using the tables from Meyer and Mayer, but acutal performance will always be poorer than predicted for clutter.

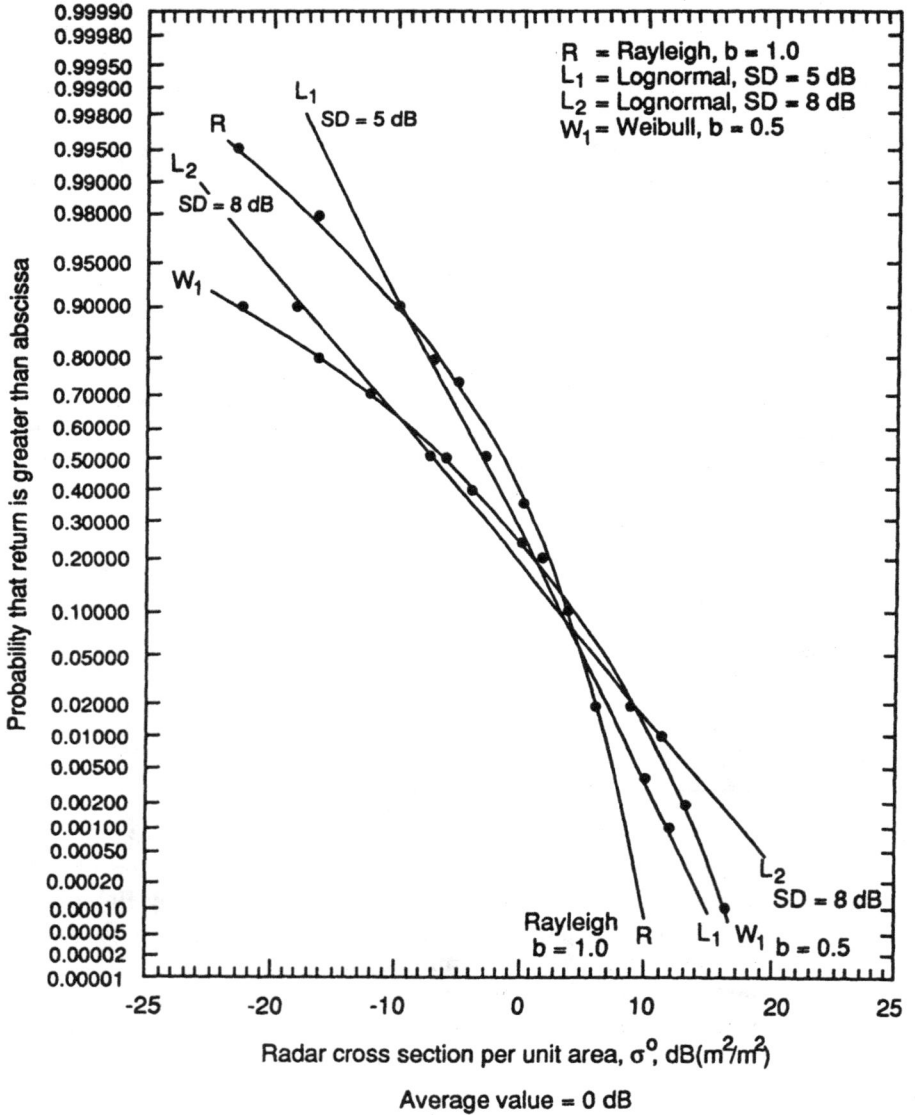

Figure 4.8 Comparison of a Rayleigh, Weibull, and lognormal distributions with the same average value (plotted as 0 dB).

4.2.2 Moving Target Processing

4.2.2.1 MTI Processing

MTI processing provides the operator with a radar system output which ideally includes only targets that are moving, eliminating stationary targets and clutter from the display. Two kinds of MTI processes are employed, noncoherent MTI and coherent MTI. Noncoherent MTI is sometimes termed *clutter-referenced MTI* because it uses the stationary clutter as the phase reference for the detection of moving targets. Coherent MTI requires coherence between the transmitter and receiver in the radar system. However, it allows for detection of moving targets in a clutter-free environment, and often has much better performance than a noncoherent MTI.

In noncoherent systems, the phase of the received vector is a randomly varying quantity, providing no information concerning the actual target phase. The phase of the vector for a coherent system provides much additional information about the target. For example, slight motion of the target will cause a pulse-to-pulse variation of the phase because of the fact that there will be a change in the radial component of range to the target. In general, radar systems do not measure the absolute range to the target in terms of wavelengths; however, a change in the number of fractional wavelengths on a pulse-to-pulse basis can be detected. The fineness with which this change can be detected is a function of the stability of the reference oscillators. If the radar oscillators exhibit phase variations, the target detected signal will change phase, even though the target may not be moving.

Change in the phase shift is exploited in several ways. In cases in which it is desired merely to determine whether a target is moving, an MTI processor may be employed. This process is essentially a high-pass filter, which, for a pulsed system, has a repeating frequency response. Targets exhibiting a Doppler characteristic will pass the filter, and targets with no Doppler, such as stationary targets and ground clutter, will not pass the filter.

Most MTI systems take the form of a single delay, double delay, or double delay feedback canceller. The delay can be implemented in one of several ways. In the early implementations of MTI processors, the delay was implemented using a quartz delay line. The delay was made precisely equivalent to the interpulse period so that the input to the subtractor network would be the response from two consecutive pulses. Since those early days, digital technology has been used to perform the delaying as well as the cancellation (subtraction) function.

If there is any target motion, the two consecutive signals will be different due to the change in phase, and there will be a resulting signal out of the subtractor. The output is full-wave rectified (because the difference may be plus or minus) and applied to a display device or signal processor. Stationary targets and stationary clutter are "canceled" and not applied to the display. Moving targets pass the canceller and show up on the display. If there is any target motion, the two consecutive signals will be different, because of the change in phase, and there will be a resulting signal out of the subtractor.

Obviously, there is no abrupt threshold between the declaration of moving and nonmoving targets. The frequency response of the canceller (filter) is a function of how many pulses are used in the cancellation process and the feedback configuration of the canceller. The system designer needs to know the spectrum of the clutter, the spectrum of the phase instabilities in the radar system, and the expected range of Doppler frequencies from the target to optimize the MTI performance of the radar system.

Several considerations need to be addressed here to complete even a cursory discussion concerning MTI processing. First, phase and amplitude instabilities in the radar itself will make stationary targets appear to be moving when processed in such a fashion. It is therefore important to minimize the phase and amplitude instabilities and design the filter to optimize the target-to-residue performance around the existent phase noise.

Secondly, it is important to point out that since the radar senses the radial component of velocity, the Doppler effects of a moving target will be reduced by a factor related to the cosine of the off-radial angle. Additionally, the radar cannot differentiate target motion from platform motion. Therefore, if the radar itself is moving, stationary targets and clutter will exhibit Doppler effects and appear to be moving. If this is to be avoided, motion compensation needs to be employed. That is, a phase correction is required before the signal is applied to the MTI canceller. The phase correction needed on a pulse-to-pulse basis is related to the platform velocity; both the speed and direction relative to the radar antenna pointing angle are considered.

4.2.2.2 Doppler Processing

An extension of MTI processing involves processing of the Doppler frequency exhibited by a moving target. The Doppler frequency is the difference between the transmitted frequency and the received frequency of a signal reflected from a moving target. The effect is essentially the same as is experienced with a train whistle as the train passes by. The apparent frequency of the whistle is higher than the actual frequency as the train approaches, and the apparent frequency is lower as it departs. The amount of Doppler frequency offset is dependent on the speed of the train. Likewise, the frequency of the wave reflected from a moving target is shifted by an amount related to the radial component of velocity, as given in Equation (4.14).

$$f_d = 2 V / \lambda \tag{4.14}$$

where

f_d = Doppler frequency
V = target velocity relative to the radar
λ = transmitted wavelength

Processing of the received signal to exploit the Doppler characteristics involves a spectral analysis of the sampled signal, where the sampling occurs at the PRF. In modern radar systems, this analysis is done by applying the fast Fourier transform (FFT) technique to the sampled and digitized data. As with the coherent MTI system, the analysis of the Doppler characteristics depends on coherence of the radar itself. Figure 4.9 shows the frequency response of a typical bank of Doppler filters implemented by means of an eight-point FFT process. The number of filters is exactly the same as the number of FFT points processed, and the total passband goes from dc to the PRF.

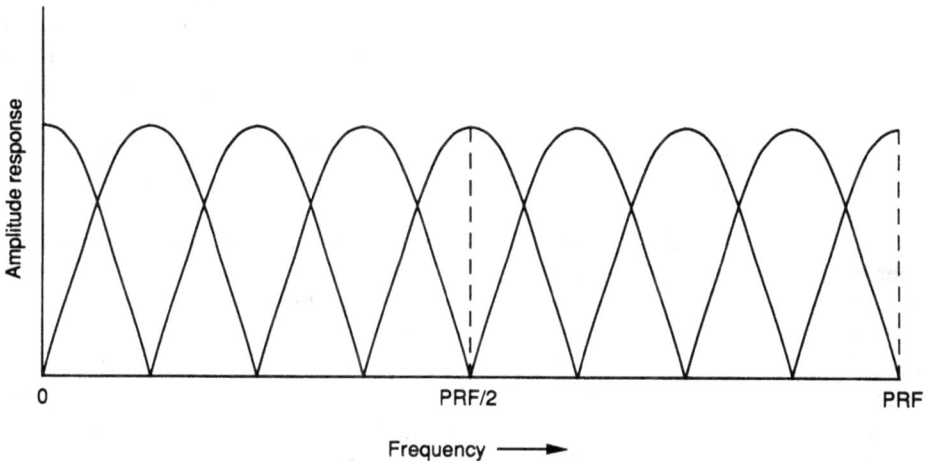

Figure 4.9 Illustration of an eight-point FFT Doppler filter bank.

The FFT performs coherent integration of the received signal. No matter what the Doppler frequency is, the signal will integrate up in one of the Doppler bins associated with the spectral analysis performed by the FFT. This is different from simple noncoherent integration from the point of view that if the signal is not of fixed phase on a pulse-to-pulse basis, the signal will not integrate up monotonically.

In other words, if the **N** vectors are not all in the same direction, when they are added, the result will not be N times the original vector. The FFT analysis can be thought of as an addition of a series of **N** vectors for a variety of pulse-to-pulse phase corrections, one of which will be correct for optimum summation. Each FFT or Doppler bin represents a different pulse-to-pulse phase correction. The bin in which the target integrates to the highest value represents the bin of the Doppler frequency due to the target velocity.

It is appropriate at this time to discuss some of the features of Doppler processing of the received signal. Since the received signal is sampled at the PRF, it is treated as a sampled data system. Most digital signal processing techniques involve Fourier transform analysis, in which case several characteristics apply. For example, the frequency resolution, F_{res}, is as fine as that allowed by the dwell time, T_d, as determined by

$$F_{res} = 1/T_d \qquad (4.15)$$

which, for a scanning antenna system, can be found by Equation (4.12).

For example, for a system with a 1° beamwidth and a scanning rate of 60° per second, the dwell time is 16.6 ms, resulting in a Doppler resolution of 60 Hz. At 95 GHz this represents a velocity resolution of about 0.09 m/s. This fine velocity resolution is important in applications involving target identification in which the target Doppler signatures are applied to pattern recognition algorithms. By contrast, an X-band system having the same dwell time or coherence time would have a velocity resolution of about 1 m/s.

Sometimes the maximum processing time is not limited by the dwell time but rather by the period of time over which the received signal is coherent, limited by either the stability of the radar system components or by the target motion. Target velocity manifests itself as a Doppler frequency or pulse-to-pulse phase shift of the received signal. Doppler processing, typically in the form of an FFT, provides a means of analyzing the spectral characteristics of the target. As long as the target velocity is constant, the components will fall into their respective bins. If the target is experiencing an acceleration, the Doppler will smear across several bins, the extent of which depends on the acceleration and the processing interval.

The Doppler resolution is found from Equation (4.15), and the Doppler spread δf_d is found from the spread in target velocity δV relative to the radar

$$\delta f_d = 2 \, \delta V/\lambda \qquad (4.16)$$

Equating the two and solving for T_d defines the limit on the processing time as a function of wavelength and acceleration by

$$T_d = \sqrt{\frac{\lambda}{2a}} \qquad (4.17)$$

For example, an acceleration, a, of 1 g (9.8 m/s^2) will result in the target Doppler slewing through about six Doppler bins during one processing period in the preceding example, in which the Doppler resolution is 60 Hz.

It is often difficult to avoid Doppler and range ambiguities with a millimeter-wave system. It can be shown that the relationship between unambiguous range, R_u, and unambiguous velocity, V_u, is

$$R_u V_u = c\lambda/8 = 118.5 \times 10^3 \text{ at 95 GHz} \qquad (4.18)$$

This means that for a system having a PRF consistent with a 5-km range the maximum unambiguous velocity is about 23.7 m/s. This is fine for many ground targets but not for most airborne targets.

4.2.3 Stationary Target Detection Techniques

For many stationary target detection problems, amplitude processing alone will not solve the detection problem because the clutter amplitude may often be equal to or greater than the return from the target. If the target is stationary, then MTI techniques cannot be used either. For these cases additional signal processing techniques are required. In general, these techniques fall into two classes: image processing and polarimetric processing. Each class of techniques can be used separately or together to achieve the desired performance. However, each technique has several drawbacks, including increased system complexity, increased dwell time on a potential target, and increased signal processing requirements.

4.2.3.1 Image Processing

Radar image processing techniques rely on obtaining high-resolution data (resolving a target into two or more resolution cells) on a target in either the range or cross-range dimension. Down-range high resolution can be obtained by transmitting a short pulse, using pulse compression, or transmitting a wide bandwidth on a pulse-to-pulse basis, or with an FMCW system. High resolution in the cross-range dimension can be obtained by using a large antenna aperture, using antenna beam sharpening techniques, or synthetic aperture radar (SAR) processing. See [10] and [11] for more information on high-resolution radar processing.

Target classification and/or identification is performed by using high-resolution target data at the frequency of interest for many aspects around the target in azimuth and elevation to "train" algorithms to recognize or classify targets. The goal is to find a "robust" algorithm that works over a large range of target orientations. Typically, the algorithms will use various spatial distributions in the target data as discriminants to eliminate more random clutter. An excellent discussion of such techniques is given in [12].

4.2.3.2 Polarimetric Processing

Polarimetric processing uses the vector nature of the scattering from targets and clutter as a discriminant. As discussed in Chapter 1, the reflectivity from a target or clutter can be completely described by the polarization scattering matrix, a 2×2 complex-valued matrix that describes the amplitude and phase characteristics of a reflected RF signal. An example of a scattering matrix in terms of vertical and horizontal polarizations is

$$\begin{vmatrix} E_{hh} E_{hv} \\ E_{vh} E_{vv} \end{vmatrix} \tag{4.19}$$

where each E_{ij} is a vector composed of an amplitude and a phase term, and represents the return from a target for transmitting polarization i and receiving polarization j.

Three types of polarimetric processing have been tried: scalar, vector, and matrix. Scalar processing involves only the amplitudes or phases of the polarization matrix. A number of scalar techniques have been evaluated, including comparison of orthogonal polarization amplitudes ($|E_{vv}/E_{vh}|$ or $|E_{vv}/E_{hh}|$) and use of the phase between two orthogonal polarizations (ϕ_{vv}/ϕ_{hh}) as a discriminant. Vector processing uses only part of the complete matrix as a discriminant but involves the use of both amplitude and phase. The dot product technique is an example in which the discriminant is of the form

$$\sigma = |E_{vv}| |E_{hh}| \cos \phi \tag{4.20}$$

where ϕ is the phase difference between E_{vv} and E_{hh}. The scalar and vector techniques are simpler to implement in terms of hardware, but have not been proven to be universally effective for target/clutter discrimination.

The final polarization processing technique involves using the entire polarization scattering matrix as a discriminant. Current techniques include using the Jones matrix, a 2×2 vector matrix, the Stokes matrix, a 1×4 power matrix, and the Mueller matrix, a 4×4 scalar matrix. The Mueller matrix can be reduced to five independent parameters of the eight total scalar parameters representing physically relevant parameters that describe a target. Setting a simple threshold for each of these parameters yields a simple but powerful discriminant. However, in order to use this technique, databases of the scattering matrices for targets and clutter are required to develop the Mueller matrix parameter thresholds, and the radar signal processor must be capable of rapidly computing these discriminants from incoming data in real time. Again, [12] includes a brief discussion of these techniques. Further discussion here is beyond the intent of this book.

4.3 EXAMPLE OF MILLIMETER-WAVE DETECTION PROBLEMS

In this section, three examples of design problems are presented for the millimeter waveband: an airborne surface search, a helicopter fire-control system, and a land-based sea search scenario. Each example summarizes the thought process that would be followed when developing the parameters for a millimeter-wave system designed to fulfill the requirements, and then evaluates the design in terms of the detectability of the desired target in clutter. Emphasis is placed on the detection problem as opposed to acquisition, tracking, and so on.

4.3.1 Millimeter-Wave Airborne Search Radar

Let us consider a scenario where it is desirable to fly a 95-GHz radar at 2-km altitude above the local terrain, as shown in Figure 4.10, so that the ground surface between 3 km and 5 km in front of the platform is under surveillance. The range of look-down angles will be from 34° to 22° (requires 12° angular coverage). The radar range will vary from 3.6 km (3-km ground distance) to 5.4 km (5-km ground distance). The spatial resolution of the system is controlled by the antenna size and the pulse length.

The antenna size is limited by space on the airborne platform. The space restricts the antenna size to less than 10 inches (25.4 cm). A 9.33 inch (23.7-cm) antenna operating at 95 GHz will produce a 1° azimuth beamwidth with the side lobes down 27 dB (Eq. (4.10)). The vertical beam for surveillance of 12° and the azimuth beam of 1° leads to a simple antenna concept of a slotted array antenna. The dimensions of WR-10 waveguide are 0.45 × 0.33 cm, and thus a vertical stacked array of six or seven slotted waveguide radiators 23.7 cm long will produce the desired beamwidths. This size is small enough to permit raster scan mechanical scan rates of 10 Hz with reasonable size drive motors. The resulting 20-Hz update rate (scans back and forth twice in 0.1 sec) will produce a smooth "picture" of the area under surveillance for $\tau = 100$ ns if the speed of the platform is less than 124 m/s. We determine this by calculating the maximum speed for which the center of the beam for the highest depression angle moves one half of the projected pulse length in 0.1 sec (1/10 Hz). The pulse length in meters is given by

$$d = c\tau/2 \text{ (m)} = 3 \times 10^8 \times 100 \times 10^{-9} \times 1/2 = 15\text{m} \qquad (4.21)$$

The pulse length projected onto the ground at the highest angle of coverage (34°) is

$$d_g = d \cos \phi = 15\text{m}/\cos 34° = 18.1\text{m} \qquad (4.22)$$

In order to have 50% overlap for a single range gate, the center of the beam must move no more than 9m in 0.05 sec or 180 m/s (648 km/hr). The speed can be

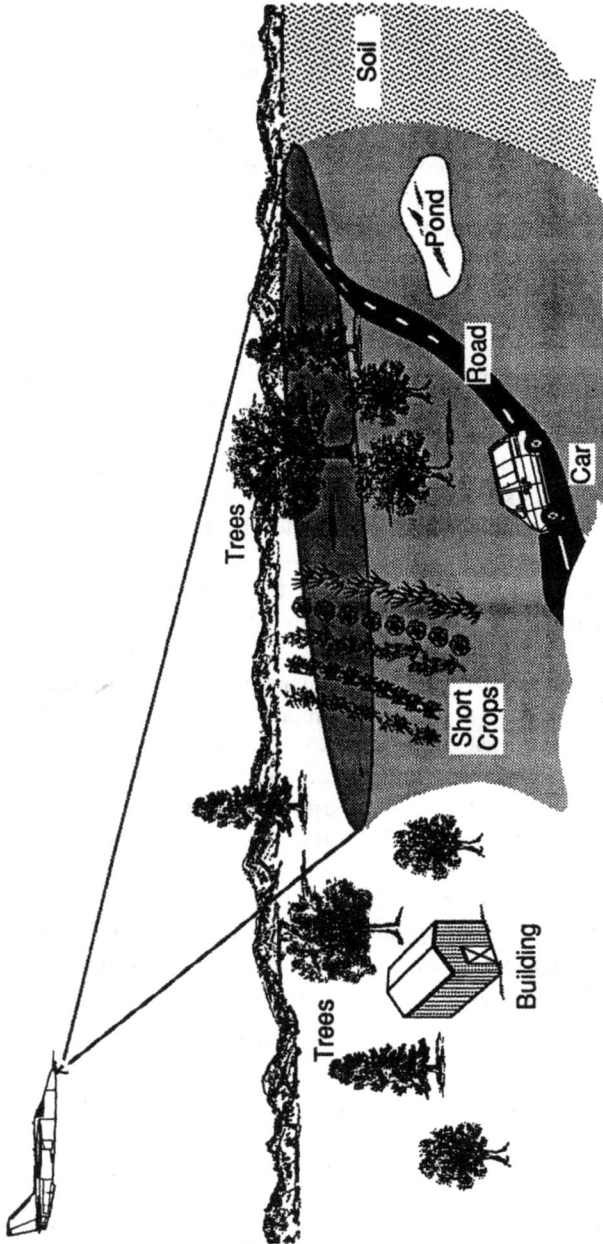

Figure 4.10 Airborne millimeter-wave surface search radar scenario.

increased by adding additional range gates to increase the amount of overlap from scan to scan.

The RCS of the terrain is given by

$$\sigma = \sigma^0 \times \text{area} = \sigma^0 \, \frac{c\tau R \theta_{az}}{2 \cos \varphi} \tag{4.23}$$

We use the system parameters for Equation (4.23) of $\tau = 100$ ns and $\theta = 1°$.

At the near ground range of 3 km, $R = 3600$ m and $\varphi = 34°$, giving

$$\sigma = -25 + 30.5 = +5.5 \text{ dBm}^2 \text{ (lakes, ponds, roads)} \tag{4.24}$$

$$\sigma = (-20 \text{ to } -12) + 30.5 = 10.5 \text{ to } 18.5 \text{ dBm}^2 \text{ (soil and seedlings)} \tag{4.25}$$

$$\sigma = (-13 \text{ to } -9) + 30.5 = 17.4 \text{ to } 21.5 \text{ dBm}^2 \text{ (crops)} \tag{4.26}$$

$$\sigma = (-10 \text{ to } -5) + 30.5 = 20.5 \text{ to } 25.5 \text{ dBm}^2 \text{ (trees)} \tag{4.27}$$

$$\sigma = (-5 \text{ to } +5) + 30.5 = 25.5 \text{ to } 35.5 \text{ dBm}^2 \text{ (urban)} \tag{4.28}$$

At the far ground range of 5 km, $R = 5{,}385$ m and $\varphi = 21.8°$, giving

$$\sigma = -25 + 31.8 = +6.8 \text{ dBm}^2 \text{ (lakes, ponds, roads)} \tag{4.29}$$

$$\sigma = (-20 \text{ to } -12) + 31.8 = 11.8 \text{ to } 19.8 \text{ dBm}^2 \text{ (soil and seedlings)} \tag{4.30}$$

$$\sigma = (-13 \text{ to } -9) + 31.8 = 18.8 \text{ to } 22.8 \text{ dBm}^2 \text{ (crops)} \tag{4.31}$$

$$\sigma = (-10 \text{ to } -5) + 31.8 = 21.8 \text{ to } 26.8 \text{ dBm}^2 \text{ (trees)} \tag{4.32}$$

$$\sigma = (-5 \text{ to } +5) + 31.8 = 26.8 \text{ to } 36.8 \text{ dBm}^2 \text{ (urban)} \tag{4.33}$$

Over this scene of 2 km down range, it is expected that ground "clutter" targets will vary in amplitude from $+5$ dBm^2 to $+37$ dBm^2.

If it is desirable to detect an automobile (typical cross section of 10 to 20 dBm^2) in this scene, then it appears that it would be difficult based on amplitude alone. There are choices to be made. Reducing the pulse length will help to reduce the "clutter," but this is a linear process. Reducing the pulse length to 50 ns will reduce the "clutter" by 3 dB, and this would help with regard to ponds, seedlings, and short crops, but not enough reduction to guarantee detection against trees and urban areas. Also, reducing the pulsewidth would decrease the overlap of the beam between scans requiring more range gates or a slower platform.

Perhaps the simplest thing to do is to drop in altitude to 400 m, which will give grazing angles from 8° to 3.4° and thus reduce the radar backscatter values for soil, short crops, and tree foliage about 20 dB. This would permit detection of a 10- to 20-dBm^2 target against all "clutter" targets except large urban areas and perhaps tree trunks. This problem is further compounded by the non-Rayleigh nature of the targets and clutter. Remember, for a reasonable P_{FA} (10^{-4}) and P_D (0.9), 12-dB signal-to-clutter ratio is required for a nonvarying target and Rayleigh clutter. However, for a time-varying target (Swerling 1) and lognormal clutter, 20-dB signal-to-clutter ratio would be required to achieve the same detection performance. Further target detection improvements will have to be produced in the frequency (Doppler) domain, the polarization domain, or using super-fine resolution techniques.

The next item to consider is the noise level and whether the targets and clutter can be detected at these ranges. Using the radar range equation

$$P_r = \frac{P_t G^2 \lambda^2 \sigma}{(4\pi)^3 R^4 L_r L_t (L_a R)(L_r R)}$$

(4.34)

Typical components available at 95 GHz are

P_t = 10 kW
L_t = 3 dB
L_r = 3 dB
L_a = 0.2 dB one way
$\log(L_r)$ = 1.6 $r^{0.64}$ dB/km one way (r = rain rate)

and from the values of $\theta = 1°$ and $\varphi = 12°$, then using Equation (4.11), $G = 34$ dB. At the maximum range of 5335 m for the radar at 400 m altitude, $P_n = -122$ dBW.

The noise level with a matched receiver is given by Equation 4.2, repeated here.

$$P_n = kTF_n Bn$$

(4.35)

For the system parameters given and a noise figure (F_n) of 5 dB, $P_N = -129$ dBW. Thus, the noise is 7 dB below the average signal level of a 10-m^2 target on a clear day. The ponds, lakes, and roads will be just at the noise level, and other "clutter" will be detected.

Consider now that a ground fog has developed. The ground visibility is 700 ft (213 m) and the fog height is 100 m. The Eldridge law shows that there is about 0.37 gm/m^3 of water, and at 10°C there would be about 0.16 dB/km of attenuation at 95 GHz. The slant range through the 100 m fog is 1247 m for our grazing angle of 4.6° at a ground range of 5000 m. The two-way attenuation will be increased by another 0.4 dB. This is not serious, and detection of the automobile and ground clutter should still be acceptable.

Instead of a ground fog, consider that a light rain of 2 mm/hr (this rate occurs about 1% of the time in central Europe and the continental U.S. [13]) is occurring

along the entire path from the radar to the 5-km ground point. The two-way attenuation coefficient of this rain rate is 5 dB/km.

At a radar range of 3000 m, the 10-m^2 target will be only 3 dB above the
noise, and detection would be marginal on a single pulse. With a scan rate of 10
Hz and a scan width of ±60°, the antenna will "dwell" on the 1° cell of resolution
for 0.28 ms.

$$T_D = T_s/S_w = (1/360 \times 10)(1/1) = 0.28 \text{ ms} \qquad (4.36)$$

where

T_D = dwell time
T_s = total one-way scan time
S_w = scan width in degrees

With 15,000 pulses per second transmitted, there would be about 4.2 pulses on the
azimuth cell of resolution every sweep. Observing the decorrelation time of wind-
blown trees in wind speeds of 5 to 6 mph, it was found that decorrelation occurs in
9 to 10 ms. Thus, the four pulses on each scan are dependent, and independent
samples will occur on a scan-to-scan basis for clutter. The four pulses will give a
3-dB ($\sqrt{4}$) improvement in the signal-to-noise case.

The revisit time T_r to a particular azimuth resolution is given by the antenna
scan rate (10 Hz) and the field of view (120°)

$$T_r = 2 \times 120/360 \times 10 = 1/15 = 67 \text{ ms}$$

If the speed of the platform is such that it moves a beamwidth in 67 ms, then some
technique will be required to place consecutive observations in a given cell of resolution to obtain scan-to-scan coverage. This can be done if there are inputs to the
radar from the platform velocity vector.

The 15,000 PPS transmitter rate will limit the radar to a 10-km range, assuming
detection of targets in the unambiguous range (first time around). This range limi-
tation due to pulse rate is acceptable since rain attenuation will be the limiting factor
of signal strength without regard to any number of pulses.

Attenuation by rain is noted to limit the range of detection. There is another
consideration of rain to be evaluated. We must determine the backscatter which will
produce an interference and possibly "mask" the target. The RCS of rain is given
by

$$\sigma = \eta \times \text{volume} = \eta \times \frac{\pi c \tau}{8} R^2 \theta_{az} \theta_{el} \qquad (4.37)$$

where

$$\eta = 0.18\, f^4 R^{1.1} 10^{-12} \ (\text{m}^2/\text{m}^3)$$

Thus, we have

$$
\begin{aligned}
f &= 95 \text{ GHz} \\
\tau &= 100 \text{ ns} \\
\text{range} &= 3000 \text{ m} \\
\text{rain rate} &= 2 \text{ mm/hr} \\
\theta_{az} &= 1° \\
\theta_{el} &= 12°
\end{aligned}
$$

giving

$$\eta = 3.14 \times 10^{-5} \text{ m}^2/\text{m}^3$$

and

$$\text{volume} = 1.95 \times 10^5 \text{ m}^3$$

Thus,

$$\sigma = +6.12 \text{ m}^2 = +7.9 \text{ dBm}^2$$

This value of σ is only 2 dB below the radar cross section of the automobile, and thus the backscatter from the rain will make it difficult to detect the automobile by amplitude only when using linear polarization.

From the data presented in Chapter 3, the backscatter from rain for circular polarization is some 10 to 15 dB less than the magnitudes of rain backscatter for linear polarization. It should also be noted that the backscatter from vehicles is typically some 5 dB less for circular polarization as compared to linear polarization. Thus, it is expected that the signal-to-clutter ratio for the vehicle when using circular polarization might be 5 to 10 dB higher in rain. This difference in signal level should offer an adequate detection margin.

When the scene moves to winter conditions with snow on the ground, then clutter conditions become more complex. Again, considering the depression angle near 30°, it can be observed in Figure 3.48 that the backscatter coefficient for refrozen snow at 95 GHz is in the range of -5 to $+5$ dBm^2. This is similar to the values for tree trunks and urban areas. On a cold winter night, the detection of vehicles against frozen snow-covered ground will be difficult. During the day, when there is thermal heating from the sun, the backscatter coefficient drops to between

−12 and −8 dBm. This value of backscatter is similar to that of crops, and detection of the vehicle should again be feasible. Another characteristic of refrozen snow is important; the variance of the return in the spatial domain is high as a result of the multiple layers resulting from the melting and refreezing process. This wide variance appears to the radar as a glinting surface. The data given in Chapter 3 indicate that integration of the return from snow over a wide bandwidth will permit fixed-point objects to be enhanced relative to the wide variations in cell-to-cell sampling of the complex frozen snow surface. Fixed frequency integration will have little effect and targets will be virtually undetectable during the refrozen phases of the snow cover.

4.3.2 Fire Control Radar for Helicopter

It is desirable to use a 35-GHz radar as an acquisition and fire control system for helicopters flying at 300 m above the local terrain with the mission of detecting military vehicles, such as tanks or armored personnel carriers (APC). Trees, ground crops, and plowed ground are found to occur in the scene and are distributed as shown in Figure 4.11, with values of σ^0 from −22 dB (m²/m²) to −8 dB. The average RCS for such a military vehicle might be on the order of +10 dBm².

The system parameters chosen are summarized in Table 4.4 and are typical of readily available technology at K_a-band. Using the radar range equation, the received power is calculated and is shown in Figure 4.12 for each type of clutter.

The RCS of the clutter increases as the grazing angle increases. At the same time, the received power of the clutter also increases with decreasing range (R^{-3}) for fixed clutter σ^0. Thus, the magnitude of clutter return does not follow a simple R^{-4} law as would a point target, but changes more rapidly with decreasing range than a point target. A 10-m² target will produce a return well above crops and plowed ground at ranges beyond 600 m; but at 600 m the tree returns are within 5 dB of the desired target return, and thus detection of the target will be marginal, remembering that for even Rayleigh clutter and a constant amplitude target, at least 12-dB SNR is required for a P_{FA} of 10^{-4} and a P_D of 0.9.

If Doppler filtering can be used to reduce the tree clutter, then the 10-dBm² objects can easily be identified. From data in Chapter 3, the frequency spectrum of tree return is given by a Gaussian plus a Lorentzian, as illustrated in Figure 4.13. At 15-mph wind speed, the equation is

$$P(f) = A e^{-1/2(f/3.4)^2} + \frac{53 \times 0.031623\,A}{1 + (f/53)^2} \tag{4.38}$$

Integrating this function from − infinity to + infinity, and setting the resulting integrated area to σ^0 (−20 dB), the value of A is found from

$$\sigma^0 = 13.155\,A \tag{4.39}$$

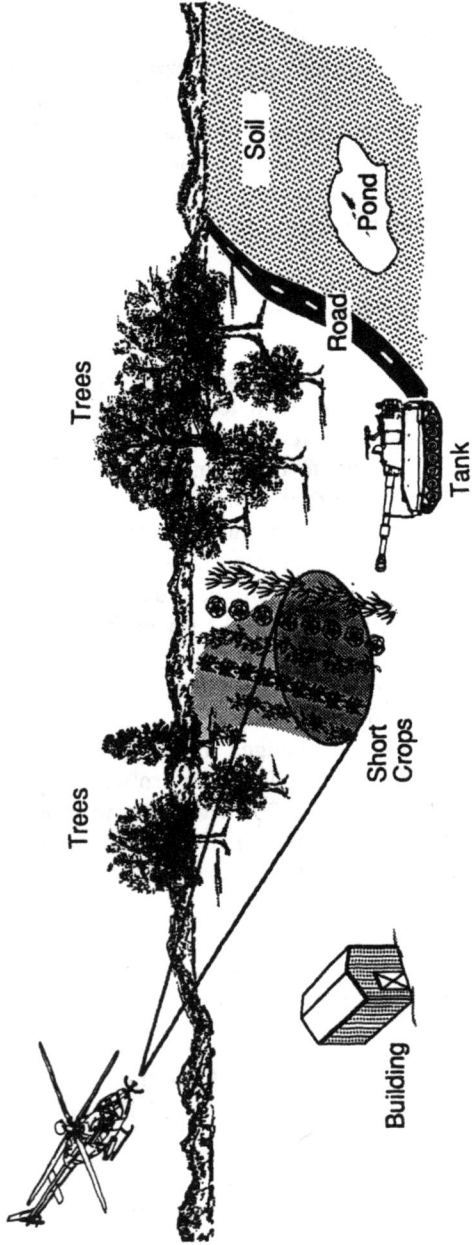

Figure 4.11 Scenario for helicopter-mounted ground vehicle fire control radar.

Table 4.4
Postulated K_a-Band Radar System Parameters

Transmitter power	=	10 kw
Transmitter pulse	=	100 ns
Antenna size	=	76 cm wide by 7.6 cm high
Antenna beam	=	0.84° by 10°
Waveguide losses	=	5 dB, receive and transmit
Clear air loss	=	0.07 dB/km one way
Noise figure	=	6 dB

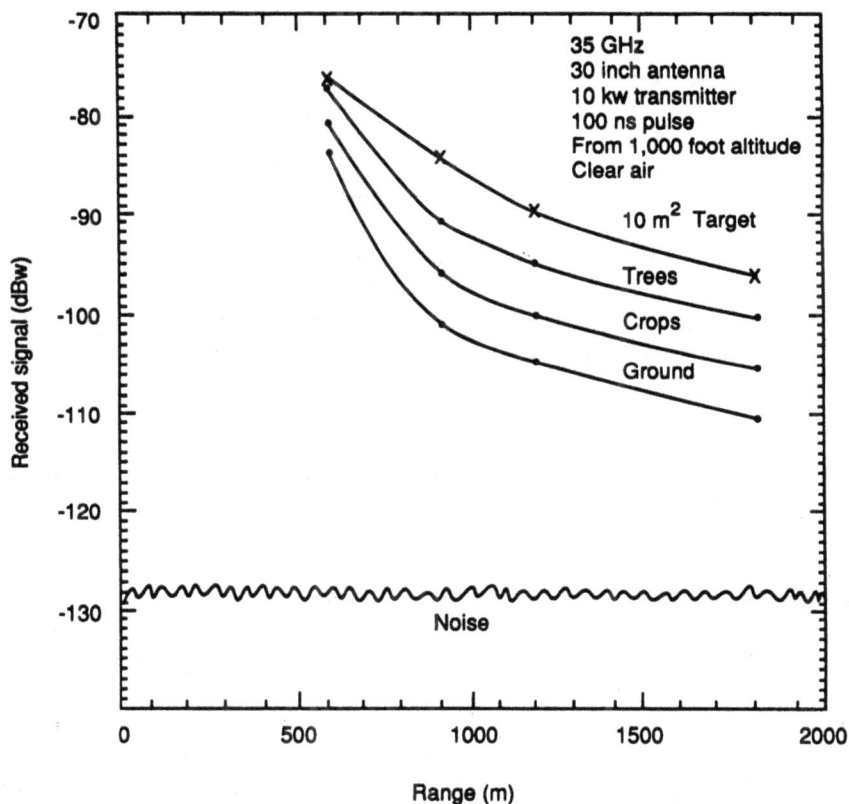

Figure 4.12 Received power versus range for the 35-GHz radar for a 10-dbm² target, trees, crops, and bare ground.

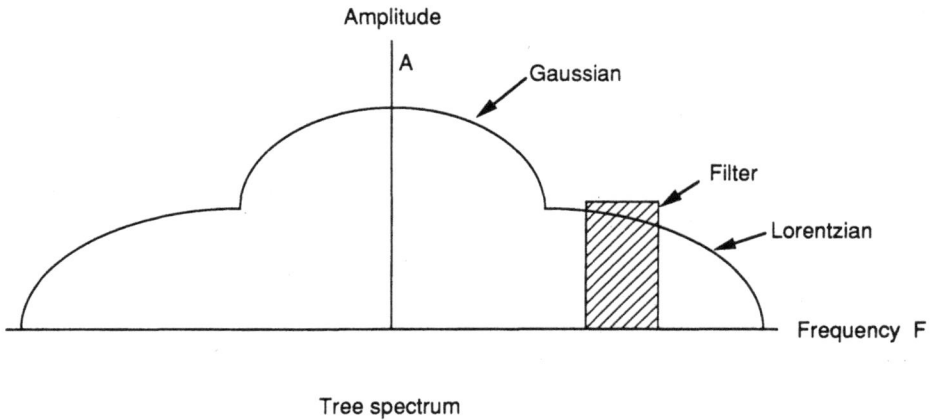

Tree spectrum

Figure 4.13 Illustration of placement of the Doppler filter for detection of the moving vehicle relative to the tree spectrum.

and

$$A = -31.2 \text{ dB } (m^2/m^2)/Hz \qquad (4.40)$$

If the velocity of the object is above 10 mph, the magnitude of $P(f)$ is down some 39 dB below the value of $P(f)$ at $f = 0$ Hz. If we wish to detect a tank moving at 2 to 3 mph (0.4 to 0.8 m/s), then we use the Doppler Equation (4.14) to calculate the frequency at which the target will occur at 35 GHz. Thus, for 2 mph,

$$f_D = 2 V/\lambda = (2 \times 0.88 \text{ m/s})/0.0086 \text{ m} = 204 \text{ Hz} \qquad (4.41)$$

When a notch bandpass Doppler filter is used from 100 to 300 Hz in order to detect objects moving at 2 mph, the total area of the power spectrum is reduced from 13.155 A to 0.5237 A. This corresponds to a 14-dB reduction in the return from tree clutter, giving an overall SNR of 19 dB, which is acceptable even for non-Rayleigh clutter and a Swerling 1 target, as discussed above. Thus, the use of Doppler processing has improved our detection situation significantly. If the radar is noncoherent and, instead of Doppler processing, clutter-referenced MTI is used, the improvement will be less than 10 dB for a typical scenario.

One concern with implementing the Doppler processor above would be having adequate dwell time on a target to provide enough data to perform the filtering. The Doppler resolution is equal to 1/(dwell time), so for a resolution of 100 Hz, which is the minimum required to separate the target Doppler from the low frequency clutter, a dwell time of 20 ms is required. For a 5,000 Hz prf, 250 pulses would occur within the dwell time so that a 256-point FFT would suffice to provide good results.

4.3.3 Sea Surface Search Radar

It is desirble to detect small craft and trash along an area of beach front in order to keep the water clear for swimmers, as shown in Figure 4.14. The radars will be mounted on small towers or small hills some 15 m above the water surface. If the craft move at moderate speeds (10 kn or 5 m/s), then Doppler processing can be used to extract the target signals from the sea clutter even though the RCS of small boats can be very low (-10 dBm^2). The Doppler frequency of the small boat will be well above the spectrum of the sea return and rain return, with the solution straightforward and with excellent Doppler detection. However, when the craft is moving at slow speeds, say 1 kt, then the Doppler frequency from the boat will be below the amplitude of the frequency spectrum of the sea return and the rain return, making it very difficult to detect the object. Let us consider two radars, one operating at 9.4 GHz and the other operating at 95 GHz.

Using curved earth geometry, the look-down angle, φ, is given by

$$\sin \varphi = \frac{H}{R} - \frac{R}{2 \, r_0} \tag{4.42}$$

where

H = height
R = range
r_0 = effective earth radius = 8500 km

From the height given, the range to the horizon is 13 km. An antenna with a vertical beamwidth of 2°, and centered at a look-down angle of 1°, will be capable of covering the surface from 400 m to the horizon. The antenna beamwidth is determined from

$$\theta = 75 \frac{\lambda}{h} \tag{4.43}$$

where

θ = half-power beamwidth, in degrees
λ = wavelength of radar signal
h = vertical dimension of antenna

Thus, the X-band antenna would be about 4 ft (1.15 m), whereas the M-band antenna would be about 5 inches (0.12 m). The drive requirements to scan a 4-ft antenna will be much greater than those for a 5-inch antenna, and the resulting cost must be considered.

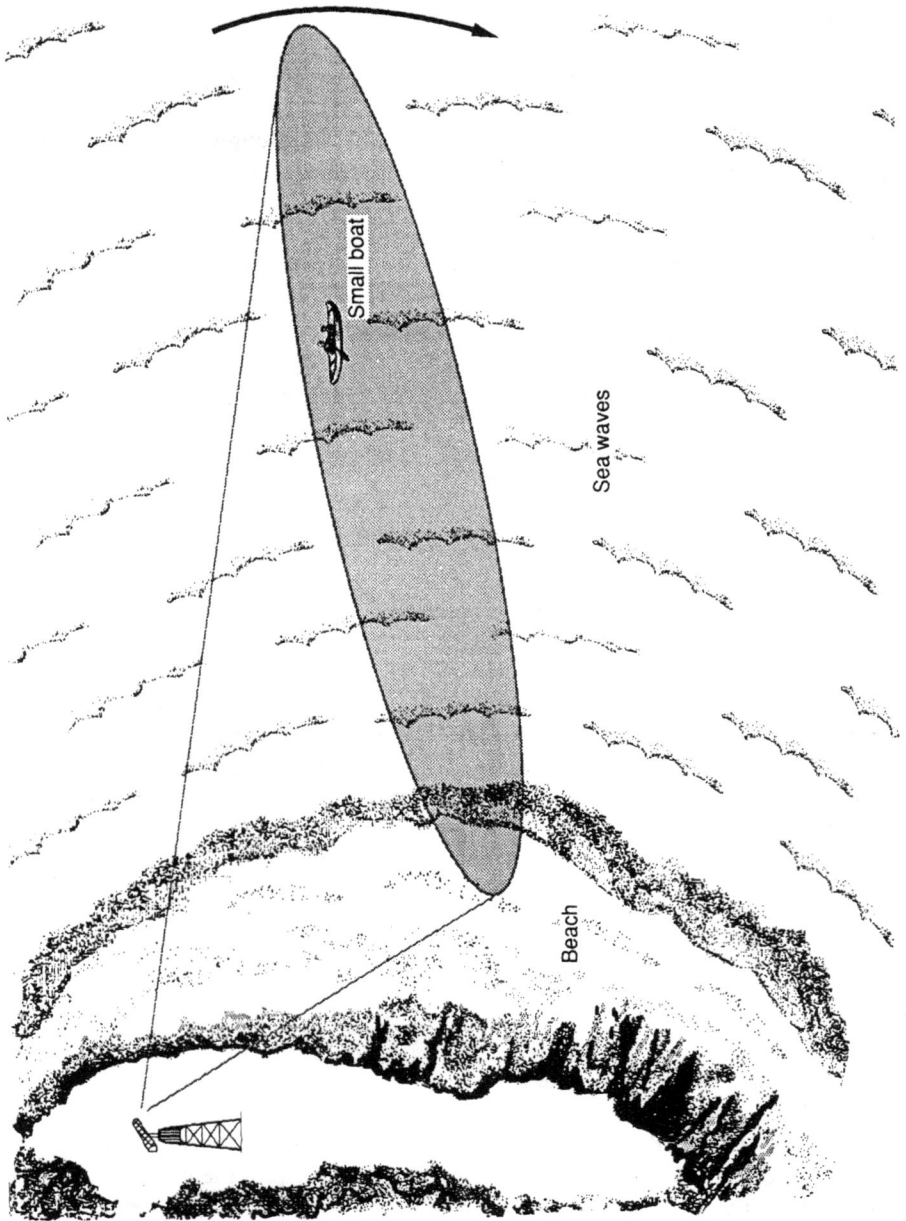

Figure 4.14 Scenario for ground-based sea surface search radar.

Now let us turn our attention to the requirements for the MTI processor and the impact of the spectrum of the sea backscatter on a clear day, and the combined sea and moderate rain spectrum on a rainy day.

Sea backscatter data were collected by Georgia Tech [14] in a multifrequency experiment that included both X-band and M-band data, and they are presented in Chapter 3. The spectrum of the sea backscatter can be modeled at X-band as

$$P(f) = A'e^{-(f/0.5513)^2} + \frac{0.1\,A'}{1 + \left(\dfrac{2\pi f}{20\pi}\right)^2} \qquad (4.44)$$

By integrating the normal distribution function,

$$\int_{-\infty}^{\infty} A'e^{1/2(x)^2}\,dx = \sqrt{2\pi}\,A's \qquad (4.45)$$

and subtracting the area under the curve when x is greater than the value of A equal to $0.1\sqrt{2}$. This is done by using the tables for the normal distribution.

In this case,

$$\text{area} = 3.0361\,A' \qquad (4.46)$$

And then by adding the area under the Lorentzian function,

$$\int_{-\infty}^{\infty} \frac{0.1\,A'}{1 + y^2}\,dy = f_c\pi(0.1\,A') \qquad (4.47)$$

and subtracting the area under the curve when y is less than the value when $A = 0.1$. This area is given by

$$\text{subtract area} = 2f_c(0.1\,A')[\tan^{-1}y]_0^{y'} \qquad (4.48)$$

where $y' = f_1/f_2$. For the values given, the total area under the Lorentzian is 2.5388 A'.

Thus, the total area under the power density spectrum of sea backscatter is 5.575 A'. This is equal to the σ^0 as measured from amplitude analysis. The value of σ^0 from the same set of data reported by Trebits and Perry [14] is given as -52 dB (m^2/m^2).

From this value of σ^0 and the area calculated, then

$$A' = -59.46 \text{ dB (m}^2/\text{m}^2)/\text{Hz} \qquad (4.49)$$

The Doppler frequency is given by

$$f_d = \frac{2v}{\lambda} \tag{4.50}$$

At X-band, 1-kt velocity will produce a frequency of 32 Hz, and 2 kt will produce a frequency of 64 Hz. Set the Doppler processor to cover the band from 30 to 100 Hz. The sea spectrum in the band pass is found from

$$\int_{f=30}^{f=100} \frac{0.1\, A' f_c}{1 + (y)^2}\, dy = f_c\, 9.1\, A' \left[\tan \frac{f}{f_c} \right]_{30}^{100} \tag{4.51}$$

$$= 0.2208\, A' \tag{4.52}$$

$$= -66\, \text{dB} \tag{4.53}$$

This is a 14-dB improvement over the -52-dB return for sea spectrum without any selective Doppler filtering.

Using typical system parameters such as pulse length = 250 ns and half power azimuth beamwidth = $2°$, the radar footprint, cell of resolution, on the sea surface at 100 m range is given by

$$\text{area} = 2R\, \frac{c\tau}{2}\, \tan \left(\frac{\theta}{2} \right) = 1309\ \text{m}^2 = 31.2\ \text{dBm}^2 \tag{4.54}$$

The unfiltered X-band RCS of the sea surface at 1000 m in a sea state of 2/3 would thus be $-52 + 31.2 = -20.8\ \text{dBm}^2$, whereas the Doppler filtered signal would have a value of $-66 + 31.2 = -34.8\ \text{dBm}^2$. These are both low values of cross section and demonstrates that the sea return should not prevent the detection of a small (1 m^2) target when above the system noise.

As stated earlier, the $2°$ beam required a 4-ft parabola at X-band, and this might require too much drive power to scan, or too large a tower for mounting, or any of many other reasons that might be undesirable. Now let us consider a radar operating at 95 GHz.

In the power spectrum domain, the sea radar backscatter coefficient is

$$\sigma^0 = A' e^{-(f/0.4934)^2} + \frac{0.01585\, A'}{1 + \left(\dfrac{2\,\pi f}{200\,\pi} \right)^2} \tag{4.55}$$

Again, it is necessary to integrate the Gaussian and the Lorentzian portions of the spectrum, subtract the overlapping areas to find the total area under the spectrum, and equate this to the average cross section per unit area as measured by Trebits and Perry. Using the same technique as was used for X-band previously, the area under the Gaussian out to 3.5 Hz is 2.7437 A', and the area under the Lorentzian from 3.5 Hz out to infinity is 4.8685 A'.

$$\text{Total area} = 7.6122 \qquad (4.56)$$

$$A' = -37 \text{ dB } (\text{m}^2/\text{m}^2) \qquad (4.57)$$

$$A' = -45.82 \text{ dB } (\text{m}^2/\text{m}^2)/\text{Hz} \qquad (4.58)$$

Again, use the same antenna beamwidth and pulse length as was used at X-band, and it is found that the RCS of the sea is -5.8 dBm^2. The detection of a 1-dBm^2 target when above noise would be considered marginal at only a 6-dB ratio. If a bandpass Doppler filter is set to cover 300 to 700 Hz, restricting detection to a 1- to 2-kt moving object, then the area of the spectrum in the filter is again given by

$$f_c \, 0.01585 \, A' \left[\tan^{-1} \frac{f}{f_c} \right]_{300}^{700} \qquad (4.59)$$

The value of $f_c = 100$ Hz from data presented in Chapter 3, and thus the value of the backscatter coefficient inside the Doppler filter, is

$$\text{area} = 0.2851 \, A' \qquad (4.60)$$

$$\sigma^0 = -51.26 \text{ dB} \qquad (4.61)$$

There is a little over 14 dB reduction of the unit area sea backscatter (from -37 dB to -51.26 dB) when passing the signal through the Doppler filter. This would give -20 dB for the sea backscatter and will produce ample target-to-clutter ratio for detection.

Having calculated that slow-moving objects at 1000m can be detected by Doppler filtering on the sea backscatter for moderate sea states of 2 or 3, let us consider that a moderate rain of 5 mm/hr enters the field of view. Rain power spectral density data were presented in Chapter 3. It was shown that the spectrum function is described at 5 mm/hr as

$$f(y) = \frac{1}{1 + (y)^3}, \, y = \frac{f}{f_c}, f_c = 27.5 \text{ Hz} \qquad \text{X-Band} \qquad (4.62)$$

$$f(y) = \frac{1}{1 + (y)^2}, \, y = \frac{f}{f_c}, f_c = 180 \text{ Hz} \qquad 95 \text{ GHz} \qquad (4.63)$$

The volume backscatter coefficient was given in Chapter 3 as

$$\eta = f^4 R^{1.7} \, 10^{-12} \, \mathrm{m^2/m^3} \qquad \text{X-band} \qquad (4.64)$$

$$\eta = 0.18 f^4 R^{1.1} \, 10^{-12} \, \mathrm{m^2/m^3} \qquad \text{95 GHz} \qquad (4.65)$$

The technique to be used to determine the amount of the rain spectrum that will appear in the Doppler bandpass filter will be the same as the technique employed to determine the amount of the sea backscatter that appeared in the Doppler filter. That is, determine the total area under the spectral density function and set that equal to η. Then determine the fraction of the area under the spectral function that would be in the Doppler bandpass and calculate the reduced value of η.

For X-band,

$$\int_{-\infty}^{\infty} \frac{f_c A}{1 + (y)^3} \, dy = f_c A \, \frac{4\pi}{3\sqrt{3}} \qquad (4.66)$$

and

$$\int_{y_1}^{y_2} \frac{f_c A}{1 + (y)^3} \, dy = f_c A \left[\frac{1}{6} \ln \frac{(1 + y)^2}{y^2 - y + 1} + \frac{1}{\sqrt{3}} \tan^{-1} \frac{2y - 1}{\sqrt{3}} \right]_{y_1}^{y_2} \qquad (4.67)$$

where $y_1 = 30/27.5$ and $y_2 = 100/27.5$.
At 9.4 GHz, the values become

$$A = \frac{3\sqrt{3} f^4 R^{1.7} \, 10^{-12}}{4\pi f_c} = \frac{3\sqrt{3}(9.4)^4 5^{1.7} \, 10^{-12}}{4\pi(27.5)} \qquad (4.68)$$

$$A = (1.811) \, 10^{-9} \, \mathrm{m^{-1}/Hz} \qquad (4.69)$$

The total $\sigma_v = (1.204) \, 10^{-7} \, \mathrm{m^{-1}}$.

After integrating $f_c A \displaystyle\int \frac{1}{1 + y^3} \, dy$ for the filter limits, we obtain $\eta = 1.4611 \times 10^{-8} \, \mathrm{m^{-1}}$ in the filter. The Doppler filter has reduced the rain return 9.16 dB.

Since the origin of the rain return and the sea return are independent sources, the total value of radar backscatter will be the sum of the sea and rain backscatter. The rain backscatter value is dependent on the volume of that portion of the antenna beam above the sea surface at the 1000 m range observation and on the pulse length.

Since the antenna has been pointed at the 1000 m range point on the sea surface, only half of the beam contains rain. This produces a volume

$$V = \frac{1}{2}\left(\frac{c\tau}{2}\right)\frac{\pi}{2}\left[R\tan\frac{\theta_{az}}{2}\right]\left[R\tan\frac{\theta_{el}}{2}\right] \tag{4.70}$$

For the system parameters used before, namely, $\theta_{az} = 2°$ $\theta_{el} = 2°$, $\tau = 250$ ns, and $R = 1,000$ m, the volume becomes 8973 m^3 and equals 39.5 dBm^3. The rain backscatter magnitude is obtained from $\sigma = \eta \times V$.

At X-band the values of rain backscatter cross section becomes

$$\sigma = -78.35 + 39.5 = -38.85 \text{ dBm}^2 \text{ in the filter} \tag{4.71}$$

and

$$\sigma = -69.19 + 39.5 = -29.69 \text{ dBm}^2 \text{ with no filtering} \tag{4.72}$$

The total backscatter from the rain and the sea at X-band becomes

$$\sigma = -33.38 \text{ dBm}^2 \text{ in the Doppler filter} \tag{4.73}$$

and

$$\sigma = -20.27 \text{ dBm}^2 \text{ with no filter} \tag{4.74}$$

These calculations show that an object of 1 m^2 should be detected at a 1000 m range by an X-band radar on the sea surface in a sea state of 2/3 and with a rain rate of 5 mm/hr.

Now let us evaluate the expected backscatter when employing a 95-GHz radar. Integrating the rain power spectral density over the limits gives

$$\int_{-\infty}^{\infty} \frac{f_cA}{1 + y^2}\, dy = f_cA\pi = \eta = (0.18)(95)^5(5)^{1.1}\, 10^{-12} \tag{4.75}$$

$$\eta = (8.611)\, 10^{-5} \text{ m}^{-1} = -40.65 \text{ dB m}^2/\text{m}^3 \tag{4.76}$$

and thus,

$$A = (1.52)\, 10^{-7} = -68.2 \text{ dB m}^2/\text{m}^2/\text{Hz} \tag{4.77}$$

With the same antenna beamwidths and pulse length as were used for the X-band radar, the volume will be the same, namely, 39.5 dBm^3. The rain backscatter becomes

$$\sigma = -40.65 + 39.5 = -1.15 \text{ dBm}^2 \text{ with no filter} \tag{4.78}$$

Integration of the rain backscatter power spectrum in the bandpass filter from 300 to 700 Hz is found from

$$\int_{y_1}^{y_2} \frac{f_c A}{1 + y^2} \, dy = f_c A[\tan^{-1} y] \qquad (4.79)$$

where $y_2 = 700/180$ and $y_1 = 300/180$. Giving

$$\eta = (7.9) \, 10 - 6 = -51.02 \text{ dB/m in the filter} \qquad (4.80)$$

and

$$\sigma = -51.02 + 39.5 = -11.52 \text{ dBm}^2 \text{ in the filter} \qquad (4.81)$$

When adding the sea backscatter with the rain backscatter in the same cell or detection at 1000m, it is found that the total return becomes

$$\sigma = -10.45 \text{ dBm}^2 \text{ with a Doppler filter} \qquad (4.82)$$

$$\sigma = 0.13 \text{ dBm}^2 \text{ with no filter} \qquad (4.83)$$

The range of 1000m does not consider the complex problem of pattern-propagation factors, multipath interference, trapping of energy at the critical angle, or shadowing by sea waves.

The conclusion is that a 95-GHz bandpass Doppler filter is required to detect a slow-moving object on the sea surface in a sea state of 2/3, and when there is a moderate rain rate of 5 mm/hr, the total level of sea backscatter is -10.45 dBm^2; thus, a small object of 1 m^2 should be detected.

REFERENCES

[1]   J.A. Scheer, private communications with N.C. Currie, Marietta, Georgia, 1992.

[2]   G.W. Stimson, *Introduction to Airborne Radar,* Hughes Aircraft Company, Los Angeles, California, 1984.

[3]   G.V. Morris, *Airborne Pulse Doppler Radar,* Artech House, Inc., Norwood, Massachusetts, 1989.

[4]   J.A. Scheer, "MMW Radar Design Considerations," Chapter 13 in *Principles and Applications of Millimeter-Wave Radar,* N.C. Currie and C.E. Brown, eds., Artech House, Inc., Norwood, Massachusetts, 1987.

[5]   M.E. Beebe and R. Yoshitani, private communications with R.D. Hayes, Hughes Aircraft Company, Fullerton, California, 1976.

[6]   L.V. Blake, "Prediction of Radar Range," Chapter 2 of *Radar Handbook,* M.I. Skolnik, ed., McGraw-Hill Book Company, New York, 1970, pp. 19–25.

[7] D.P. Meyer and H.A. Mayer, *Radar Target Detection,* Academic Press, New York, 1973, pp. 156–157.

[8] D.K. Barton, *Radar Systems Analysis,* Artech House, Inc., Norwood, Massachusetts, 1979, p. 146.

[9] W.K. Rivers, "Low Angle Radar Sea Return at 3-mm Wavelength," Final Report on Contract N62269–70-C-0489, Georgia Institute of Technology, Atlanta, Georgia, 1970.

[10] J.L. Kurtz and J.A. Scheer, "High Resolution Measurements," Chapter 10 in *Radar Reflectivity Measurement: Techniques and Applications,* N.C., Currie, ed., Artech House, Inc., Norwood, Massachusetts, 1989, pp. 369–431.

[11] D.L. Mensa, *High Resolution Radar Imaging,* Artech House, Inc., Norwood, Massachusetts, 1982.

[12] W.A. Holm, "MMW Radar Signal Processing Techniques," Chapter 6 of *Principles and Applications of Millimeter-Wave Radar,* N.C. Currie and C.E. Brown, eds., Artech House, Inc., Norwood, Massachusetts, 1987, pp. 241–312.

[13] R.D. Hayes, data supplied by ETAC, Scott AFB, Illinois, 1976.

[14] R.N. Trebits and B. Perry, "Multifrequency Radar Sea Backscatter Data Reduction," Georgia Tech Project A-2717, Naval Surface Weapons Center, 1982.

Glossary

a	Radius of sphere, cylinder, and so forth
A	area
A_e	antenna effective aperture; effective earth radius
a_{hh}	horizontal transmit and receive component of the polarization matrix
a_{hv}	horizontal transmit and vertical receive component of the polarization matrix
A_i	interference factor
A_u	upwind-downwind factor
A_w	wind speed factor
APC	armored personnel carrier
a_{vh}	vertical transmit and horizontal receive component of the polarization matrix
a_{vv}	vertical transmit and receive component of the polarization matrix
b	Weibull slope parameter
B	bandwidth
B-scope	display with azimuth horizontal and range vertical
B_n	noise bandwidth
BW	radian beamwidth of an antenna
c	velocity of light (approximately 3×10^8 m/s)
CF	correlation function
CFAR	constant false alarm rate
cm	centimeter
d	diameter or distance
D	drop diameter
dB	decibel

dbh	tree trunk diameter 4.5 feet above the ground
dBm^2	decibels relative to a square meter
e	eccentricity; exponential function 1
E_h	horizontal polarized component of electric field
E_r	electric field magnitude at the radar receiver
E_i	electric field magnitude transmitted, incident at the object
E_v	vertical polarized component of electric field
f	frequency (Hz)
f_c	frequency for which a PSD is reduced to one half its zero frequency value
f_d	Doppler frequency
FFT	fast Fourier transform
FM CW	frequency modulated constant wave
F_n	noise figure
g	gram or acceleration with respect to gravity
G	antenna gain
GHz	10^9 Hz
G_r	receive antenna gain
G_t	transmit antenna gain
HC	hexachloroethane
HE	high explosives
h	hour
h_a	radar antenna height
h_d	duct height constant
H_r	magnetic field magnitude at the radar receiver
H_i	magnetic field magnitude transmitted, incident at the object
$H_n^{(1)}(ka)$	Hankel functions
Hz	Hertz (1 cycle per second)
I	inphase
$J_0(ka)$	modified bessel function of the first kind
k	wave number ($= 2\pi/\lambda$); refractive index factor ($(m^2 - 1)/(m^2 + 2)$); Boltzman's constant
km	1,000 m
kt	nautical miles per hour
L	length (target, antenna); one way loss

LL	transmit left circular, receive left circular polarization
LSF	least square fit
m	complex index of refraction of water; meter
M	liquid water content per unit volume of fog (g/m^3)
MHz	10^6 Hz
MTI	moving target indication
$N(D_i)$	drop size distribution
ns	10^{-9} sec
PPI	plan position indicator (display)
PSD	power spectral density
P_a	power density at antenna
P_r	received power
prf	pulse repetition frequency
P_D	probability of detection
P(f)	power spectral density function
P_{FA}	probability of false alarm
P_t	transmit power
P(X)	cumulative probability function
Q	quadrature
r	ellipticity ratio
r_0	effective earth radius
R	slant range; rain rate
RCS	radar cross section
RP	red phosphorus
R_e	equivalent rainfall rate
RL	transmit right circular, receive left circular polarization
RR	transmit right circular, receive right circular polarization
$R(\tau)$	autocorrelation function
R_u	unambiguous range
s	second
S	standard deviation
SAR	synthetic aperture radar
SDV	swimmer delivery vehicle
SNR	signal-to-noise ratio
t	time

T	time interval; time period; temperature, (°C)
T_d	dwell time
V	volume; target velocity relative to radar
V_u	unambiguous velocity
WP	white phosphorus
W(X)	probability density function
X	snow mass concentration; random variable
$\bar{X}$	average value of X
X_m	median value of X
$Y_n(ka)$	Weber's functions
Z	meteorological reflectivity factor, (mm^6/m^3)
α	beam shape factor (= 1.33 for Gaussian beam shape); logarithmic attenuation coefficient (dB/m); half-angle of cone; incidence angle
α'	linear attenuation coefficient (nepers/m)
Δh	rms surface roughness
γ	surface reflection factor (= $\sigma°/\sin \Theta$)
η	backscatter coefficient (m^2/m^3)
n	radar volumetric backscatter coefficient (dB m^2/m^3)
Θ	grazing angle (= $\pi/2$ incident angle)
θ_{az}	radar antenna azimuth beamwidth
θ_{el}	radar antenna elevation beamwidth
λ	radar wavelength
μm	10^{-6} m
π	an irrational number expressing the ratio of the circumference to the diameter of a circle (approximately 3.141592)
ρ	density of particles (g/cm^3)
σ	radar cross section

$\bar{\sigma}$	average radar cross section
σ°	backscatter coefficient for a surface (radar cross section per unit area) (m^2/m^2)
$\overline{\sigma^\circ}$	average backscatter coefficient
τ	radar pulse length (sec.); 1/bandwidth for pulse compression systems; ellipticity angle
ϕ	phase or phase difference
Φ	angle between boresight and upwind
Φ_x	phase of x polarization component
Φ_y	phase of y polarization component
ω	radian frequency $(= 2\pi f)$

Index

The Artech House Radar Library

David K. Barton, Series Editor

EREPS: Engineer's Refractive Effects Prediction System Software and User's Manual, developed by NOSC

High Resolution Radar, Donald R. Wehner

High Resolution Radar Cross-Section Imaging, Dean Mensa

Interference Suppression Techniques for Microwave Antennas and Transmitters, Ernest R. Freeman

Introduction to Electronic Defense Systems, Fillippo Neri

Introduction to Electronic Warfare, D. Curtis Schleher

Introduction to Sensor Systems, S.A. Hovanessian

IONOPROP: Ionospheric Propagation Assessment Software and Documentation, Herbert Hitney

Kalman-Bucy Filters, Karl Brammer

Laser Radar Systems, Albert V. Jelalian

Lidar in Turbulent Atmosphere, V.A. Banakh and V.L. Mironov

Logarithmic Amplification, Richard Smith Hughes

Machine Cryptography and Modern Cryptanalysis, Cipher A. Deavours and Louis Kruh

Millimeter-Wave Radar Clutter, Nicholas Currie, Robert Hayes, and Robert Trebits

Modern Radar System Analysis, David K. Barton

Modern Radar System Analysis Software and User's Manual, David K. Barton and William F. Barton

Monopulse Radar, A.I. Leonov and K.I. Fomichev

MTI and Pulsed Doppler Radar, D. Curtis Schleher

Multifunction Array Radar Design, Dale R. Billetter

Multisensor Data Fusion, Edward L. Waltz and James Llinas

Multiple-Target Tracking with Radar Applications, Samuel S. Blackman

Multitarget-Multisensor Tracking: Advanced Applications, Volume I, Yaakov Bar-Shalom, ed.

Multitarget-Multisensor Tracking: Advanced Applications, Volume II, Yaakov Bar-Shalom, ed.

Over-The-Horizon Radar, A.A. Kolosov, et al.

Principles and Applications of Millimeter-Wave Radar, Charles E. Brown and Nicholas C. Currie, eds.

Pulse Train Analysis Using Personal Computers, Richard G. Wiley and Michael B. Szymanski

Radar and the Atmopshere, Alfred J. Bogush, Jr.

Radar Anti-Jamming Techniques, M.V. Maksimov, et al.

Radar Cross Section Analysis and Control, A.K. Bhattacharyya and D.L. Sengupta

Radar Cross Section, Eugene F. Knott, et al.

Radar Electronic Countermeasures System Design, Richard J. Wiegand

Radar Evaluation Handbook, David K. Barton, et al.

Radar Evaluation Software, David K. Barton and William F. Barton

Radar Meteorology, Henri Sauvageot

Radar Polarimetry for Geoscience Applicatons, Fawwaz Ulaby

Radar Propagation at Low Altitudes, M.L. Meeks

Radar Reflectivity Measurement: Techniques and Applications, Nicholas C. Currie, ed.

Radar System Design and Analysis, S.A. Hovanessian

Radar Technology, Eli Brookner, ed.

Radar Vulnerability to Jamming, Robert N. Lothes, Michael B. Szymanski, and Richard G. Wiley

RGCALC: Radar Range Detection Software and User's Manual, John E. Fieldng and Gary D. Reynolds

SACALC: Signal Analysis Software and User's Guide, William T. Hardy

Secondary Surveillance Radar, Michael C. Stevens

SIGCLUT: Surface andVolumetric Clutter-to-Noise, Jammer and Target Signal-to-Noise RadarCalculation Software and User's Manual, William Skillman

Signal Detection and Estimation, Mourad Barkat

Small-Aperture Radio Direction Finding, Herndon Jenkins

Solid-State Radar Transmitters, Edward D. Ostroff, et al.

Space-Based Radar Handbook, Leopold J. Cantafio, ed.

Spaceborne Weather Radar, Robert M. Meneghini and Toshiaki Kozu

Statistical Signal Characterization, Herbert Hirsch

Statistical Theory of Extended Radar Targets, R.V. Ostrovityanov and F.A. Basalov

Surface-Based Air Defense System Analysis, Robert Macfadzean

Surface-Based Air Defense System Analysis Software and User's Manual, Robert Macfadzean and James M. Johnson

VCCALC: Vertical Coverage Plotting Software and User's Manual, John E. Fielding and Gary D. Reynolds